D1499845

Improving Competence across the Lifespan

Building Interventions Based on Theory and Research

MCP HAHNEMANN UNIVERSITY
HAHNEMANN LIBRARY

Improving Competence across the Lifespan

Building Interventions Based on Theory and Research

Edited by

Dolores Pushkar
William M. Bukowski
Alex E. Schwartzman
Dale M. Stack

and

Donna R. White

Concordia University
Montreal, Quebec, Canada

Plenum Press • New York and London

Library of Congress Cataloging-in-Publication Data

On file

BF
713
I34
1996

Proceedings of the Conference on Improving Competence across the Lifespan,
held November 15 – 17, 1996, in Montreal, Quebec, Canada

ISBN 0-306-45814-4

© 1998 Plenum Press, New York
A Division of Plenum Publishing Corporation
233 Spring Street, New York, N.Y. 10013

http://www.plenum.com

10 9 8 7 6 5 4 3 2 1

. All rights reserved   -

No part of this book may be reproduced, stored in a retrieval system, or transmitted in any form or by any
means, electronic, mechanical, photocopying, microfilming, recording, or otherwise, without written
permission from the Publisher

Printed in the United States of America

PREFACE

This book arises from a conference held in November 1996 designed to examine how competence can be improved in the different stages of the lifespan. To this end, we brought together eminent researchers in different areas of human development — infancy, childhood, and adulthood, including the late adult years. The conference was based on the premise that discussion arising from the interfaces of research and practice would increase our knowledge of and stimulate the further application of effective interventions designed to improve competence.

The editors wish to acknowledge the contributions of Concordia University and the Fonds pour la Formation de Chercheurs et l'Aide à la Recherche (FCAR) in providing funding and other assistance toward the conference "Improving Competence Across the Lifespan" and toward the publication of this book.

Finally, we wish to express our gratitude to the numerous students associated with our Centre for their help and to Gail Pitts and Lesley Husband of the Centre for Research in Human Development for their assistance. We are especially grateful to Donna Craven, Centre for Research in Human Development, for her heroic work on both the conference and the present volume. We could not have met our goals without you.

The Editors:

Dolores Pushkar
William Bukowski
Alex E. Schwartzman
Dale M. Stack
Donna R. White

September 1997

CONTENTS

Emergent Themes in Studying Competence across the Lifespan: An Introduction 1
Dolores Pushkar and Dale M. Stack

Chapter 1. Competence across the Lifespan: Lessons from Coping with Cancer 9
Bruce E. Compas and Alexandra Harding

Chapter 2. Emotional Competence and Development 27
Michael Lewis

Chapter 3. Socioemotional and Cognitive Competence in Infancy: Paradigms,
Assessment Strategies, and Implications for Intervention 37
Dale M. Stack and Diane Poulin-Dubois

Chapter 4. The Socialization of Socioemotional Competence 59
Nancy Eisenberg

Chapter 5. The Socialization of Children's Emotional and Social Behavior by
Day Care Educators 79
Donna R. White and Nina Howe

Chapter 6. Competence: A Short History of the Future of an Idea 91
William M. Bukowski, Tanya A. Bergevin, Amir G. Sabongui, and Lisa A. Serbin

Chapter 7. Adolescent Use and Abuse of Alcohol 101
Frances E. Aboud and Shelley C. Dennis

Chapter 8. Self-Regulation as a Key to Success in Life 117
Roy F. Baumeister, Karen P. Leith, Mark Muraven, and Ellen Bratslavsky

Chapter 9. Improving Competence in Prime Adulthood: Of Sows' Ears and
Pygmalions 133
Alex E. Schwartzman

Chapter 10. Responding to the Challenges of Late Life: Strategies for Maintaining
and Enhancing Competence 141
Sherrie Bieman-Copland, Ellen Bouchard Ryan, and Jane Cassano

Chapter 11. Interventions to Improve Cognitive, Emotional, and Social
            Competence in Late Maturity 159
   Dolores Pushkar and Tannis Arbuckle

Chapter 12. Political and Scientific Models of Development 177
   Arnold Sameroff and W. Todd Bartko

Afterword: Improving Competence — Is It Ethical to Consider Cost? 193
   Frederick Lowy and Anna-Beth Doyle

Index 201

EMERGENT THEMES IN STUDYING COMPETENCE ACROSS THE LIFESPAN

An Introduction

Dolores Pushkar and Dale M. Stack

Centre for Research in Human Development
Department of Psychology
Concordia University

The definition of competence that underlies the chapters in this book emphasizes the adaptive functioning of the individual in adjusting to the environment and coping with life tasks effectively and without unnecessary psychological costs. This adaptive process helps to determine the experience of normative life circumstances and becomes even more important in coping with stressful life events. Such adaptive functioning includes the emotional, cognitive and social skills necessary to live effectively and in reasonable harmony with the self and with others (Masterpasqua, 1989). This definition of competence partakes of the positive mental health approach applied to the study of adjustment (Sherman, 1993). That is, the emphasis is not just on the remittance of symptoms, but on continued development of potential so that functioning improves and more closely reflects optimal levels. For example, interventions at all stages of the lifespan past infancy are created not only to alleviate depression, but also to improve social functioning to enhance the well-being of individuals.

However, the developmental tasks that individuals confront across the lifespan stages vary. Furthermore, both social and personal resources differ, including the acquired and genetically based abilities developed by individuals. Consequently, as we move from childhood to late maturity there is change, as well as consistency, and increasing heterogeneity in the types of competencies required, developed and maintained. Similarly, across the lifespan different types of interventions are required and have varying degrees of success in assisting individuals in functioning more competently. The focus of this book in considering the theories of development and the success of applications fostering competence across all stages of the lifespan is thus, unique.

People live in a "world that is complex, multifaceted, and messy" (Anderson, 1995). Developmental research provides valuable information in enabling us to attain the goals of creating more effective interventions in such a world, both in the development and testing

Improving Competence across the Lifespan, edited by Pushkar *et al.*
Plenum Press, New York, 1998.

of explanatory theories that allow us greater predictive effectiveness, and in the development of multi-measure assessments. The chapters in this volume provide a rich summary and evaluation of important theoretical positions and basic research on the different developmental stages. Similarly, the chapters provide useful evaluations of the effects of different types of interventions targeting a variety of functions in both immediate and long-term frameworks. Such approaches help to explicate the behavioral changes that are causally linked to age-based developmental processes or to social processes, and the different remedial approaches required by each. Finally, the chapters cover promising new directions for future interventions, as well as indicating those directions and areas that need further examination and development.

CHAPTER SYNOPSES

In the opening chapter, Bruce Compas and Alexandria Harding consider the importance of coping processes to achieve and maintain competence in normative contexts and in stressful situations at different points in the lifespan. Using the context of coping with a severe life threatening stress, that of cancer, they discuss how competence, and the resources and skills it requires, varies across the lifespan. Adult and child cancer patients and their families, and their responses to the threats and challenges of the illness and its ramifications, are described to illustrate how competent functioning is maintained.

Michael Lewis largely focuses on emotional development in the early years in his chapter, arguing that emotional competence, or learning how to cope with emotions, is a major developmental task. He particularly emphasizes the self-conscious emotions of pride, embarassment, shame and guilt, highlighting the importance of the socialization and utilization of these emotions. He discusses the sophistication of preschoolers in dealing with attributions of success and failure and the importance of assessing attributions to self or performance, and illustrates how this develops over time and between the sexes. He underscores the importance of the socialization experience of children and the role that it can play. In his view, to understand competence, we need to understand how the emotional life of the child develops.

Dale Stack and Diane Poulin-Dubois review the status of research in infant competence and the history of assessment and intervention as it applies to the early years. The authors focus on infant socioemotional and cognitive competence and review both existing and promising assessment tools to evaluate competence in infants in these domains. They argue that powerful paradigms and theories developed from research in infancy have important implications and applications to early intervention and have, already to date, been proven influential. The chapter covers the research related to infant assessment and the challenges researchers face, as well as, some of the research directions that are important to pursue. The importance of context and individual differences are underscored.

Nancy Eisenberg reviews the theoretical context and empirical results of studies examining parental socialization of children's socioemotional competence, especially in emotionally evocative contexts. The chapter reviews such aspects as parental responses to children's expressions of different types of emotions, for example hurtful and sorrowful emotions, and how they relate to children's responses to peers. The chapter also examines family risk factors on children's development such as poverty and family conflict.

The influence of day care educators on the socialization of children's emotional and social behavior is examined by Donna White and Nina Howe. Their chapter reviews studies of the influence of home care versus day care and quality of child care environments on the development of social and emotional competence in preschool and school aged children. Using the parent socialization literature, including the work of Nancy Eisenberg, they develop and examine a model of teacher-child interaction and its influence on children's development. Their chapter also considers factors in child care environments that facilitate social and emotional development in relation to possible interventions.

William Bukowski, Tanya Bergevin, Amir Sabongi, and Lisa Serbin examine the historical development of the concept of competence, tracing the influence of the theorizing of White, Piaget and Erikson on present day research. The central aspects of competence as a dynamic interactional process is examined in a discussion of adolescent experience and peer relations. The authors use multiple levels of analysis examining the relations between competence within the context of the individual, the dyad and the group.

Frances Aboud and Shelley Dennis survey one of the problem areas in adolescent socialization of competence, adolescent abuse of alcohol. Their chapter examines the effects of such variables as ethnicity, age, gender and cohort. The authors specify risk factors and examine the effectiveness of preventive interventions that attempt to reduce alcohol abuse and so prevent or reduce the sequelae of negative social, emotional and academic functioning.

The chapter by Alex Schwartzman discusses the concept of competence as applied to developmental processes in the adult years, focussing on the developmental tasks of establishing and maintaining a family and the acquisition of work skills enabling economic sufficiency. The chapter examines transitional periods in adulthood as providing opportunities for the remediation of maladaptive behaviors through interventions which minimize the constraining effects of biological processes and prior negative learning experiences. The design and delivery of interventions relating to work skills and socioemotional competence in low income individuals are examined.

The chapter by Roy Baumeister, Karen Leith, Mark Muraven and Ellen Bratslavsky posits self-regulation, the ability to alter one's response, as a distinctive and potent ability in humans that influences successful functioning. Self-regulation, which includes affect regulation, attentional control and impulse regulation, is examined in relation to aspects of competent functioning that are essential for independent adult lifestyles, such as goal setting and reaching behavior, procrastination and the avoidance of self-destructive behavior. The influence of self-esteem and environmental context on self-regulation is discussed.

Sherrie Bieman-Copland, Ellen Bouchard Ryan and Jane Cassano focus on competence in late maturity. Their chapter presents an extension of the Baltes and Baltes model (1990) of selection, optimization and compensation by considering internal factors and external factors, such as stereotyped expectancies. Such factors influence the perception of older adults' competence and beliefs about their ability to cope with environmental demands. The chapter assesses the importance of social stereotypes as determinants of competent functioning in elderly samples.

Dolores Pushkar and Tannis Arbuckle evaluate the success of various interventions designed to improve competent functioning in late life. They review age-related functioning and problems in cognition, emotion and social behavior and examine the potential of the various implications of specific theories and research findings for the development of interventions.

The influence of political models on human development are discussed by Arnold Sameroff and W. Todd Bartko, who examine issues arising out of scientists' and practitioners' concerns for human development and politicians' responsibilities for the making and funding of policies which shape the availability and nature of interventions. Sameroff and Bartko illustrate the role of scientists in creating empirical questions out of different political models in the context of their study investigating environmental influences on cognitive and social-emotional development from infancy to adolescence. Their research illustrates the effects of ecological risk factors in the child's world working with Bronfenbrenner's (1979) microsystems.

In an afterword, Frederick Lowy and Anna-Beth Doyle discuss the difficult decisions to be made concerning the ethics of resource allocation in societies faced with increasing demands for service and limited resources. The authors consider the ethical issues arising from different types of interventions, those designed to achieve or restore normative levels of function or those designed to increase competence beyond the normative levels that would otherwise be attained. The influence of different contexts on such outcomes is also examined. The authors examine three models of the decision-making processes involved in differentiating the necessary from the merely desirable in the allocation of resources for the improvement of competencies.

EMERGENT THEMES

Despite the diversity of subject populations, problems and methodologies, some common issues and themes emerge in the chapters with regard to the creation and application of interventions. The theme of individual differences and the history of the individual across his or her lifespan was raised in various ways in nearly all of the chapters. The theme of mediating influences of genetic or dispositional variables on environmental contexts is discussed by most chapter authors (see for example the chapters by Lewis and by Schwartzman). The possible constraints imposed by such genetic processes involved in individual differences are found at all stages of the human lifespan. Rather than just discussing the separation of environmental and genetic contributions to individual development, the contributions of both to individual differences were emphasized by the chapter authors. Increasing attentional focus on the changing course of development might make an approach relating genetic and environmental interactions more appealing (Overton, 1997). Such a changing course could also take us back (or forward depending on how you look at it) to the single case in some instances, which as Valsiner (1997) in his address on developmental ideas in the 20th century indicates, is the "definitive empirical source for data derivation." In his words, "aggregate knowledge about how the system works is based on one person and extended to new selected single cases" (Valsiner, 1997). For practitioners faced with the needs of individual people, this remains the prime directive.

The fact that individual differences is a common theme is not surprising, given the importance of considering the individual in applied work, the fact that development is different for different people, and the emphasis that individual differences receive in the current literature. Making the history of the individual more salient promotes an individually tailored approach to intervention or at least a consideration of individual factors within an otherwise group approach. Furthermore, since research studies leave much unexplained variance, individual case and life histories can often help to elucidate factors that have played important roles. However, focusing on individual differences solely as *error* is an adevelopmental view (Horowitz, 1990). Moreover, the populations with whom our chapter authors work are heterogeneous, thus it is almost impossible to produce optimal results in applying assessment or intervention strategies to the entire group without taking some account of the individual (Bieman-Copland, Ryan, & Cassano; Bukowski, Sabongui, Bergevin & Serbin; Pushkar & Arbuckle; Schwartzman; Stack & Poulin-Dubois). The latter point is aptly illustrated in some of the chapters in what may be referred to as "client effects," that have been relatively unconsidered in determining intervention outcomes. These may be due to inadvertent factors, which require accommodation on the part of practitioners, such as expectations about therapy that reflect different cohort socialization and experiences. The objectives of the researcher or practitioner are not necessarily shared by the people for whom the intervention is designed. The chapters by Sherrie Bieman-Copland, Ellen Bouchard Ryan and Jane Cassano, and Dolores Pushkar and Tannis Arbuckle, report research findings indicating that while clients might learn more effective ways of functioning, they do not necessarily generalize such techniques to other contexts or use them in the near future. The roles of such factors as effort and fatigue in influencing clients to function at a level of competence that is a compromise between increased effectiveness, comfort, and familiarity, deserve further research.

Furthermore, interventions can have inadvertent effects which, although unknown to the practitioner, influence response to the intervention. For example, clients, particularly older people and children, usually receive a great deal of support from their social networks of family and friends. Such informal social support can enable an individual to function at a higher level than could be predicted on the basis of assessment of individual skills alone. But interventions can have effects which spread to and alter the functioning of the social context of the client and hence, the client's response. Such spreading effects of interventions and the effect of feedback from network members probably are not known to the researcher or practitioner. Caution in taking a sufficiently broad history and assessing the life circumstances of the client is necessary.

A lifespan developmental perspective (Dixon & Lerner, 1988) and changes in the traditional views of development are underscored in the chapters. According to Fogel, Lyra and Valsiner (1997) a radical change in world view has infiltrated the traditional social sciences, a world view that is based on such concepts as complexity and chaos, for example. This change in thinking and world view is beginning to take hold and has come about because of the underlying theme of dynamic development and the recognition of the role of perspectives which study dynamic self-organizing systems. William Bukowski and colleagues, for example, highlight the conceptual advances that have been recently made in our understanding of competence. However, as Alex Schwartzman emphasizes this change has not equally affected thinking in all areas of lifespan development. Although theoretical development in areas of child and late maturity has been greatly advanced, the theoretical understanding of developmental processes in the mid-adult years lags behind. The developmental perspective, in Valsiner's (1997) view, is one that embraces "becoming" and focuses on relative stability and change. As Ellen Ryan and her colleagues emphasize, in late maturity even to remain stable requires innovation.

Such theoretical perspectives raise the possibility that developmental outcomes and trajectories may not be "predictable in any simple, linear or causal way from their antecedents" (Fogel et al., 1997). Moreover, from a pragmatic perspective we can pit thoughts of indeterminism with the contention that some determinism in development may be accounted for by political and social forces which operate to permit or remove opportunities for services, as Arnold Sameroff and W. Todd Bartko, and Frederick Lowy and Anna Beth Doyle indicate. Take, for example, the early intervention movement and the contemporary focus on service for all preschoolers in prevention efforts. We know that in actuality this is not what is taking place. Similarly, with increases in medical and research technology and knowledge, life spans for the elderly are longer. Consequently, the quality of life in the later years is an area requiring more attention, but the soaring costs of health care combined with budget cuts ensure that many elderly people do not get the help they need. The shift from home to group care as discussed by Donna White and Nina Howe also reflects changes in policy and costs.

In addition, there seems to be a concerted attempt to account for variance in stabilities in competence, along with a guiding assumption that prediction is the direction of choice for developmental researchers and practitioners. However, instability may be an index of dynamic change over time (Horowitz, 1990). "An alternative or companion strategy is to focus on documenting the range and stability and variability of individual differences in young infants, and to study how these individual differences interact with environmental conditions" (Horowitz, 1990, p. 15). Ultimately, in Horowitz's view, the goal is to understand continuities and discontinuities in development, not as products, but as process. More process oriented strategies and questions move us closer to asking how individual differences figure in developmental processes. It is not merely a predictive power issue (can we predict to later development) that we grapple with as developmentalists nor whether we can rightfully discriminate between groups on some construct or task, but rather, we study the process of development over real time as individuals interact with environments. This notion of individual-in-context surfaced as a salient issue in a number of the chapters (for example, Compas & Harding; Stack & Poulin-Dubois; Aboud & Dennis). Development just like knowledge must be situated (Overton, 1997).

The inter-relations among developmental domains is a related and important issue and recognition of the inter-relationships is necessary to inform plans for intervention. As Dale Stack and Diane Poulin-Dubois contend in their chapter, problems or referral issues in one domain may carry over into other domains or have implications for other domains. As they describe, such an interface holds particularly true for cognition and emotion in infancy but also in adulthood as emphasized by Roy Baumeister and his colleagues. However, what these interfaces imply, in general, is that we need measures which are comprehensive and which assess aspects of more than one domain, i.e., comprehensive assessment strategies (Achenbach, 1982). A multi-domain, multi-contextual approach where information is collected in multiple settings over several time points and a number of response domains is recommended (e.g., Cicchetti & Wagner, 1990; Achenbach, 1982). Such multi-contextual assessments allow for better determinations of the contributions of constitutional, cognitive and socioemotional factors, for example (Cicchetti & Wagner, 1990). But they also reflect an assessment of the "whole" person, rather than a specific attempt at measuring one ability in a standardized setting at one point in time.

A comprehensive approach to assessment applies equally well to intervention. In addition, a serious emphasis on prevention is required. Here, the systematic study of risk and protective factors as potential precursors of dysfunction or health is important (Coie et al., 1993). The study of at-risk groups serves an important role in helping us to better understand development, developmental change and the course of aberrant development. The burgeoning field of developmental psychopathology reflects this focus (e.g., Burack & Enns, 1997; Lewis & Miller, 1990). Arnold Sameroff and W. Todd Bartko and Dale Stack and Diane Poulin-Dubois illustrate the value of promising research paradigms and their application to risk groups during infancy. A life span approach is critical in such applications. For example, the salience of risk factors may fluctuate developmentally, and there may be cumulative effects of exposure to many risk factors (Coie et al., 1993). Behavior and environmental events may have quite different meaning and significance at different ages. Moreover, vulnerability to disorders may vary with time and vary in intensity. Individual characteristics and attributes of the environment may thus be considered to be general types of protective factors (Coie et al., 1993).

Developmental models must thus move away from simple prediction and emphasize the complex transactions between individuals, environments and time (Coie et al., 1993; Fogel et al., 1997; and as illustrated in the chapter by Sameroff and Bartko). This is an area where longitudinal, prospective studies, as well as, prospective, intergenerational studies of risk and non-risk groups provide a valuable means of advancing our knowledge and providing important clues to intervention. They also permit an important opportunity to evaluate the role of cohorts, and the salient aspects of the history of the individual's life. Moreover, these approaches allow for the monitoring of developmental trajectories over time, and the explication of, or at least elucidation of, moderating and mediating variables. However, lifespan measures need to be developed so that the ways in which interactions are revealed or the way in which a specific behavior changes over the life cycle can be observed. In addition, as Coie et al. (1993) advocate, an increased emphasis on creating "contextually and culturally sensitive measures" is important. Finally, an increasing focus on including the multiple systems of influence (e.g., family, school, peer, work, etc.) under which human behavior occurs across the life span is essential to our understanding and to enhance competence.

Conceptualizing development as it occurs within multiple systems of influence (Bronfenbrenner, 1979; see also Sameroff & Bartko) is important and such conceptualizations naturally bring to the foreground the socialization process. The role of socialization is not solely played out during childhood but is also salient during infancy, adulthood and older adulthood. The specific socialization force or "socializer" (parents, peers, teachers, work place, friends and family, health care system) may vary and the extent of the influence may be to a lesser or greater degree. However, its influence is evident. Moreover, these are bidirectional influences where the individual interacts with the environment and the socialization process. Given that there are both external and internal influences on development, how socialization plays a role at the different ages and life stages becomes intriguing. Some compelling examples include the socialization of emotion (for example the chapters by Eisenberg and by Lewis) and the socialization process as it applies to the development of a social network as discussed by William Bukowski and colleagues and Frances Aboud and Shelley Dennis. In addition, regulation can be both social and self-regulation and self-regulation came through the chapters as a central theme (as illustrated by Nancy Eisenberg and Roy Baumeister and colleagues in this volume, but also by others, for example, Schore, 1994).

The influence of social expectancies are also powerful at all lifespan stages and our authors point out the dangers inherent in attributing skills deficits to internal causes and neglecting the pervasive effects of social stereotypes and expectancies specific to individuals. Sherrie Bieman-Copland, Ellen Bouchard Ryan and Jane Cassano point out the misattribution of lower levels of competence to internal factors, such as personal ability, and age limitations, rather than to external factors such as bias, stereotypes, inappropriate measures. Such influences are as likely to affect the performance of younger as that of older subjects.

Research and practice can cross-fertilize each other in different areas, providing novel and effective approaches to the study of competent functioning (see the chapter by Baumeister and colleagues). For example, Donna White and Nina Howe point out the potential utility of research findings and theory from the area of parent-child socialization to the study of teacher-child interaction in examining the effects of child care on children's competence. Similarly, the chapter by Dolores Pushkar and Tannis Arbuckle discusses minor dysphoria among elderly people and raises the possible role of inability to create a meaningful context for oneself as a precipitating factor. However, meaning requires a social structure as well as individual effort. The chapter by Frances Aboud and Shelly Dennis points out the influence of cultural factors on adolescent problematic use of alcohol. An examination of the role of dysphoria on adolescent abuse of alcohol could be beneficial. In both conditions, elderly dysphoria and adolescent alcohol use, the behavior is part of a continuum ranging from normal to problematic, with the problematic not clearly defined.

In closing, the chapters also provide examples of interventions that have been unsuccessful and explanations for their failure. It is important to recognize and differentiate failures attributable to methodology, execution or to unrealistic objectives. As Frederick Lowy and Anna-Beth Doyle emphasize, this is particularly salient in a time of widespread government cutbacks of services and costly private services. Practitioners working with troubled humans are well aware that they work with people who face hard facts and, sometimes, have only hard choices to make. One of the most difficult decisions to make is to decide not only *when* and *how* one should intervene, but also *whether* one should try to intervene.

No one would deny that prevention is the best intervention possible. In considering ways to foster competence and to prevent the development of problematic, aberrant, or less competent functioning, the most effective approach is based on a broad social policy. Approaches must also be based on complex models of human development integrating research and practice. The recent emphasis on prevention is a welcome development and if accompanied by adequate funding should benefit society and individuals.

REFERENCES

Achenbach, T. M. (1982). *Developmental psychopathology* (2nd ed.). New York: John Wiley.

Anderson, E. (1995). Knowledge, human interests, and objectivity in feminist epistemology. *Philosophical Topics, 23*, 27–58.

Baltes, P. B., & Baltes, M. M. (1990). Psychological perspectives in successful aging: The model of selective optimization with compensation. In P. B. Baltes & M. M. Baltes (Eds.), *Successful Aging: Perspectives from the Behavioral Sciences* (pp. 1–27). New York: Cambridge University Press.

Bronfenbrenner, U. (1979). *The Ecology of Human Development.* Cambridge: Harvard University Press.

Burack, J. A., & Enns, J. T. (Eds.) (1997). *Attention, development, and psychopathology.* New York: Guilford Press.

Cicchetti, D., & Wagner, S. (1990). Alternative assessment strategies for the evaluation of infants and toddlers: An organizational perspective. In S. J. Meisels & J. P. Shonkoff, (Eds.), *Handbook of Early Childhood Intervention.* New York: Cambridge University Press.

Coie, J. D., Watt, N. F., West, S. G., David Hawkins, J., Asarnow, J. R., Markman, H. J., Ramey, S. L., Shure, M. B., & Long, B. (1993). The science of prevention: A conceptual framework and some direction for a national research program. *American Psychologist, 48*, 1013–1022.

Dixon, R. A., & Lerner, R. M. (1988). A history of systems in developmental psychology. In M. H. Borstein & M. E. Lamb (Eds.), *Developmental Psychology: An advanced textbook* (2nd ed., pp. 3–50). Hillsdale, NJ: Lawrence Erlbaum.

Fogel, A., Lyra, M. C. D. P., & Valsiner, J. (1997). *Dynamics and indeterminism in developmental and social responses.* Mahwah, NJ: Lawrence Erlbaum.

Horowitz, F. D. (1990). Developmental models of individual differences. In J. Colombo & J. Fagen (Eds.), *Individual differences in infancy: Reliability, stability, prediction* (pp. 3–18). Hillsdale, NJ: Lawrence Erlbaum.

Lewis, M., & Miller, S. M. (Eds.) (1990). *Handbook of developmental psychopathology.* New York: Plenum Press.

Masterpasqua, F. (1989). A competence paradigm for psychological practice. *American Psychologist, 44*, 1366–1371.

Overton, W. F. (1997, August). Between positivism and relativism: Healing the dichotomy. Paper presented at the 105th annual convention of the American Psychological Association, Chicago, IL.

Schore, A. N. (1994). *Affect regulation and the origin of the self: The neurobiology of emotional development.* Hillsdale, NJ: Lawrence Erlbaum.

Sherman, E. (1993). Mental health and successful adaptation in later life. In M. A. Smyer (Ed.), *Mental Health and Aging* (pp. 85–92). New York: Springer Publishing.

Valsiner, J. (1997, August). Well-kept hostages: Developmental ideas in the 20th century. Paper presented at the 105th annual convention of the American Psychological Association, Chicago, IL.

1

COMPETENCE ACROSS THE LIFESPAN[*]

Lessons from Coping with Cancer

Bruce E. Compas and Alexandra Harding

University of Vermont

The characteristics that comprise competent functioning are principal elements of adaptive social, emotional, and cognitive growth and development. Moreover, aspects of competence may serve as important protective factors that reduce the risk for psychopathology and illness. The attributes that constitute competence differ, however, with development as the resources and skills that are needed for effective functioning change with the major developmental tasks and roles in childhood, adolescence, and adulthood. We will consider these issues in the context of how competence serves as a resource in adapting to a life threatening illness — cancer. Cancer patients and their families are faced with a series of stressors and adversities that challenge psychological well being and quality of life, and call upon sources of competence as a part of effective coping and adaptation. Because cancer presents children and adults with threats and challenges that are objectively relatively similar, it offers the opportunity to study competence and coping in individuals of different ages who are faced with comparable forms of stress.

Our goal is to consider developmental differences and consistencies in competence across the lifespan, using the process of coping with cancer as an opportunity to observe important aspects of competent functioning. We begin by considering definitions of competence, focusing on those aspects of competence that are reflected in the process of coping with stress. We then briefly review the stressful aspects of cancer that challenge individuals' competencies. Next we review the literatures concerned with the characteristics of coping with cancer in adults, children, and families. Finally, we offer a synthesis of the lessons we can learn about competence from the study of adaptation to cancer.

---

[*] This chapter is based on a presentation at the conference "Improving Competence Across the Lifespan," Concordia University, November, 1996.

Improving Competence across the Lifespan, edited by Pushkar *et al.*
Plenum Press, New York, 1998.

CONCEPTUALIZATIONS OF COMPETENCE

Competence has been conceptualized in terms of the features of cognitive, behavioral, and emotional functioning that comprise effective interactions with the environment. This typically includes characteristics of individuals that are related to their success in accomplishing the major developmental tasks of each point in the lifespan. For example, Masten and colleagues have identified three central developmental tasks of childhood: getting along with peers, achievement in academics, and conduct that abides by the salient rules of school and family (Masten, Coatsworth, Neeman, Gest, Tellegen, & Garmezy, 1995; Masten, Hubbard, Gest, Tellegen, Garmezy, & Ramirez, 1997). Successful achievement of these developmental tasks is reflected in separate but related indices of social, academic, and behavioral competence (Masten et al., 1995).

Competence is closely related to the concept of resilience, which refers to competent functioning in the face of significant adversity (Masten et al., in press). Competent functioning is represented by successful adaptation under normative conditions, whereas resilience is reflected in successful functioning under conditions of stress and adversity (Masten et al., 1997). Individuals who are low in competence are presumed to display poor functioning under either normative or adverse conditions, as their lack of resources for competent functioning leaves them vulnerable to poor outcomes regardless of the circumstances that they face. Highly competent individuals will achieve positive developmental outcomes under normative conditions, and will display resilience by achieving adaptive functioning in the face of adversity.

Our perspective on competence is reflected in our conceptualization of positive mental health and effective coping (Compas, 1993; Compas, Connor, Harding, Saltzman, & Wadsworth, in press). Positive mental health is defined as "a process characterized by development toward optimal current and future functioning in the capacity and motivation to cope with stress and to involve oneself in personally meaningful instrumental activities and/or interpersonal relationships. Optimal functioning is relative and depends on the goals and values of interested parties, appropriate developmental norms, and one's sociocultural group" (Compas, 1993, pp. 166–167). One central element of competence from this perspective is concerned with how the individual functions under non-normative conditions — under stress or faced with adversity. Individuals under stress must mobilize their resources, both their personal competencies and their social relationships and resources, in order to meet a significant challenge, threat, or loss. The nature of successful adaptation to stress, and the skills and resources needed to successfully cope, change with development (Compas et al., in press).

Coping as Enacted Competence

The process of coping with stress reflects competence that has been enacted, competence that has been put into action in order manage or overcome conditions of stress and adversity. This perspective draws on process models of coping, best exemplified in the work of Lazarus and Folkman (1984) who define coping as cognitive and behavioral efforts to master, manage, tolerate or reduce a problematic relationship between the person and the environment. Coping responses have been differentiated along several dimensions, including the goal or intention of the response (to solve the problem or increase personal control, or to palliate one's emotions or increase secondary control), the mode of response (cognitive or behavioral), and whether it involves engagement with or disengagement from the stressor, or one's emotional response to stress.

Coping responses are a subset of a broader category of individuals' cognitive, behavioral, affective, and physiological responses to stress that are important in understanding successful and unsuccessful adaptation. Specifically, coping refers to effortful or volitional responses but does not include involuntary or automatic reactions to stress that influence and are influenced by coping responses (Compas et al., in press; Compas, Connor, Osowiecki, & Welch, 1997). Involuntary responses are the result of individual differences in temperament and associative learning or conditioning processes during prior exposure to stress. One form of involuntary responses, the propensity to experience intrusive negative thoughts about a stressor, is particularly important in determining adaptive versus maladaptive outcomes of stress (Baum, Cohen, & Hill, 1993). Some types of coping, including temporary distraction, are useful in managing intrusive thoughts and other forms of involuntary stress responses. Other types of coping, including efforts to suppress unwanted thoughts, are ineffective in managing intrusion and other automatic responses, and may increase the severity and duration of uncontrollable negative thoughts (e.g., Wegner, 1994). Therefore, competent patterns of coping are affected by temperamental characteristics and by prior exposure to stress.

There is no simple formula that defines effective coping. Effective coping and adaptation are influenced by the type of stressor, cognitive appraisals of the stressor, and the individual's developmental level, and personal and social resources. Individuals and families who are faced with the diagnosis of a potentially life threatening illness provide vivid and compelling examples of competent and effective functioning under adversity. We now turn our attention to research on the psychosocial aspects of cancer to better understand the nature of competent functioning in the face of stress.

CANCER AS A STRESSOR

The lives of virtually everyone will be touched by cancer, either directly because they are diagnosed with the disease or indirectly because a friend or relative is diagnosed. Current prevalence rates in the United States indicate that there are over 1,000,000 new diagnoses of cancer per year, and over 500,000 deaths annually due to cancer (American Cancer Society, 1996). The prognosis and course of various types of cancer differs greatly, however. For example, breast cancer and prostate cancer are highly prevalent, affecting over 180,000 women and 200,000 men, respectively, per year. The prognoses for these forms of cancer are relatively encouraging, however, as the median 5-year survival rates are 75–80%. Lung cancer, another highly prevalent form of the disease with over 170,000 new diagnoses per year, has a median 5-year survival rate of only 13%. These statistics reflect new developments in surgical, medical, and radiologic treatments which have transformed some forms of cancer from an acute condition with very little possibility of survival to a chronic condition for many patients who live for many years after their diagnosis and treatment. Other forms of cancer continue to present an immediate threat of loss of life.

The nature of cancer as a significant stressor is best conveyed by considering case examples of patients and families coping with this disease. Susan is a 49-year-old woman who became concerned when she discovered a lump in her left breast while taking a shower. A mammogram determined that there was a 5 centimeter mass, the results of a biopsy confirmed that the lump was cancerous, and she underwent a complete mastectomy and analysis of her lymph nodes revealed that two of them contained cancer cells. She then began several weeks of radiation therapy, followed by 6 months of chemotherapy. Susan's illness presented multiple stressors for her and her family. This began with the sudden and unexpected nature of the diagnosis, which raised immediate fears for her and her family that she might die from the disease. She experienced physical pain from her surgery, fatigue as a result of her radiation and chemotherapy treatments, and she lost her hair as a consequence of chemotherapy. During her treatments and for several months after, she was unable to work full time at her job, and she was frequently too tired and ill to fulfill her usual roles at home — cooking meals, shopping, cleaning, laundry, driving her children to school and activities, and gardening. As a result, her husband and her children, especially her 14-year-old daughter, had to assume many of the responsibilities that she usually carried. Her husband became increasingly angry and withdrawn over the course of her treatment, and she noted that her daughter seemed much more emotional and easily upset. Susan was also unable to take part in the activities that typically brought so much pleasure to her and her family, including spending enjoyable time with her children reading or in other activities, and going out with her husband. In summary, it seemed that there was hardly an aspect of her life that had not been affected in some way by her experience with cancer.

Joseph is a 5-year-old boy whose development and health had been normal until sometime around his fourth birthday. At that time, he developed a persistent fever that was unaccompanied by any other flu-like symptoms, and his parents noticed that he bruised very easily. Joseph's parents brought him to his pediatrician, and the results of a preliminary blood count indicated leukemia. Joseph was then sent to a specialist who completed a bone marrow aspirate to confirm the diagnosis and subtype of leukemia, followed by a lumbar puncture to determine if the leukemia had spread to his central nervous system (CNS). Joseph was diagnosed with Acute Lymphoblastic Leukemia (ALL), and Joseph was admitted to the hospital for an initial phase of his treatment, induction chemotherapy, which eliminated the active disease, followed by intrathecal chemotherapy on an outpatient basis to prevent the occurrence of CNS disease. Finally, he was placed on maintenance chemotherapy for two years, which consisted of weekly injections at the oncology clinic. During the course of treatment, Joseph received frequent venipunctures, and intermittent bone marrow aspirates and lumbar punctures to monitor the effectiveness of treatment.

Just as Susan's illness presented multiple stressors for her and her family, Joseph's illness overwhelmed his family. The diagnosis of ALL in a previously healthy child shattered many of Joseph's parents' assumptions and raised the possibility he could die. Joseph's parents had difficulty communicating around issues of the disease, as Joseph's mother preferred not to think about the disease, while Joseph's father wanted to talk about what they were going through. Joseph's treatment and medical procedures were also very stressful for the family. He lost his hair and vomited frequently from chemotherapy. He also became so distressed by the medical procedures involving needles that he initially had to be restrained. Joseph's mother had to give up her job in order to care for Joseph, as he could not be put in a day care center in the initial phases of his treatment. Joseph's father became tearful and emotional upon visits to the hospital when Joseph had to undergo a medical procedure, and he began to have dreams at night about Joseph's treatment and procedures. Joseph's mother and father began to spend less time together alone, as most of their attention and focus was on Joseph, and they began to argue frequently when they were alone. It is obvious, as in Susan's case, that most aspects of Joseph's family life were dramatically affected by his cancer.

COPING AND COMPETENCE IN ADULT CANCER PATIENTS

The most extensive research on adults' adaptation to cancer has focused on breast cancer patients, and as a consequence we will draw extensively on that literature here.

Numerous studies have documented that the stress associated with the diagnosis and treatment of breast cancer is related to higher levels of psychological symptoms, most notably affective distress (symptoms of depression and anxiety), and decreased quality of life for women with the disease when compared to normative community samples not faced with a similar level of stress (e.g., Derogatis et al., 1983; Fallowfield, Hall, Maguire, & Baum, 1990; Hilton, 1989; see Glanz & Lerman, 1992, for review). Longitudinal analyses suggest that distress declines steadily in the months following diagnosis (Morris, Greer, & White, 1977). Research has also shown, however, that a significant subgroup of women with breast cancer manifest psychological symptoms that meet criteria for psychiatric disorder at a rate that exceeds that found in the normal population (e.g., Nelson, Friedman, Baer, Lane, & Smith, 1994; Schag et al., 1993). This subgroup appears distinct in that these women are at high risk for prolonged symptoms of affective distress, lasting for months to years after the initial diagnosis (Schag et al., 1993). Estimates of the percentage of women who comprise this high risk group range from 15% to over 40%, with the median in the range of 25% to 35% (e.g., Compas et al., 1997; Ganz et al., 1993; Nelson et al., 1994; Schag et al., 1993).

Four broad factors have been identified that characterize successful adaptation to cancer among adults. These factors provide a profile of the characteristics of competent functioning in the face of this severe stressor. The first category involves sociodemographic and disease characteristics, including the age and education of the patient, and the type and severity of cancer. The second involves enduring personal characteristics and social relationships that serve as resources for adapting to cancer, foremost of which are dispositional optimism and social support. The third category includes individual differences in involuntary cognitive, affective, and behavioral responses, especially the propensity to experience involuntary, intrusive negative thoughts about the cancer. And the fourth category involves the ways that individuals cope with the stress of cancer, including personal coping efforts and mobilization of social support. Research indicates that coping responses are influenced by the first three categories of responses, and that coping mediates the effects of these other factors on emotional adjustment.

Demographic and Disease Characteristics

Age, education, economic factors, cancer severity, degree of physical impairment, and extent of surgery have all shown some association with distress among cancer patients (e.g., Vinokur, Threatt, Vinokur-Kaplan, & Satariano, 1990; Schag et al., 1993; Stoll, Oppedisano, Epping-Jordan, Harding, Compas, & Krag, 1997). Specifically, younger patients and those with less education experience greater emotional distress. Age may be a risk factor for emotional distress among younger women with breast cancer, as the disease is more unexpected among younger women, they may experience more disruptions to the developmental tasks of young adulthood (e.g., beginning a career, decisions about childrearing, raising young children), and have fewer women in their peer group who have experienced breast cancer and can provide social support. Women with less formal education may experience greater distress because they may have less experience processing the complex information that confronts them as part of their diagnosis and treatment, and they may have less experience with health professionals as a result of more limited economic resources.

The effects of age and education on emotional adjustment appear to be mediated by the ways that women cope with the stress of breast cancer. For example, Stoll et al., (1997) found a negative correlation ($r = -.30$) between age and symptoms of anxiety and depression in breast cancer patients near the time of their diagnosis. Younger women used more emotional expression/ventilation, and this coping response was a significant mediator of the association of age and distress. This suggests that younger women may rely more than older women on at least one form of coping, the ventilation of feelings, that is detrimental in its effects. The association of education level with emotional distress is also mediated by the use of another form of emotion-focused coping, in this case the use of avoidance (Epping-Jordan et al., 1997). Women with less formal education are more likely to use avoidance coping, which in turn was related to more distress, suggesting that women with less formal education may lack the resources or the experience to use more active forms of engagement or approach coping.

Resources for Adjustment: Optimism and Social Support

The dispositional trait of optimism–pessimism has been shown to be related to psychological adjustment to breast cancer, with dispositional optimists displaying lower symptoms of anxiety and depression at the time of their diagnosis and subsequently over the next year (Carver et al., 1993, 1994; Epping-Jordan et al., 1997). Optimism is represented in positive expectations about future outcomes, in this case positive beliefs about the course of one's cancer. Optimists also report that they are able to cope by finding positive features of adverse circumstances, including a stressor as challenging as breast cancer. Optimism reduces negative emotions such as depression and anxiety, and mobilizes active, problem-oriented coping efforts.

The role of supportive, confiding relationships has been examined extensively in the adjustment of women with breast cancer and findings indicate that perceptions of the availability of supportive relationships is related to more successful psychological and physical adaptation to breast cancer (see reviews by Bloom, 1986; Nelles, McCaffrey, Blanchard, & Ruckdeschel, 1991). Women who perceive higher levels of social support in their relationships with family and friends report higher self-esteem, less psychological distress, and experience better medical outcomes, including increased natural killer cell activity and longer survival (e.g., Bloom & Spiegel, 1984; Funch & Marshall, 1983; Levy et al., 1990). Furthermore, the effects of social support on psychological and physical adjustment may be mediated, at least in part, by the types of coping strategies that women use and the source of their support (e.g., Dakof & Taylor, 1990).

Intrusive Thoughts and Avoidance

Two features of a stress response syndrome (Horowitz, 1982) or of Post Traumatic Stress Disorder (American Psychiatric Association, 1994) are important in the process of adjustment to cancer — intrusive thoughts and avoidance of these thoughts. Intrusive thoughts are uncontrollable negative images and thoughts about a stressor, and avoidance involves cognitive and behavioral efforts to suppress and avoid reminders of the stressor. These two responses are important predictors of both maladaptive emotional and health outcomes among cancer patients. Intrusive thoughts are strongly associated with heightened symptoms of anxiety and depression. In contrast, avoidance appears to be a stronger predictor of poor health outcomes than anxiety/depression and intrusive thoughts (Epping-Jordan, Compas, & Howell, 1994). Thus, in addition to being highly correlated with other symptoms of psychological distress, efforts to avoid or suppress cancer specific thoughts and emotions are a marker of poorer disease outcomes in cancer patients.

Although intrusive thoughts and avoidance are moderately correlated ($r = .50$), there may be important differences in combinations of high and low levels of these factors.

Primo, Compas, Oppedisano, Epping-Jordan, and Krag (1997) identified four groups of breast cancer patients near the time of their diagnosis based on patterns of high and low levels of intrusive thoughts and avoidance. The four groups differed in anxiety and depression symptoms at diagnosis, and 3- and 6-month follow-ups, with the most consistent differences occurring between those who were high in both avoidance and intrusive thoughts and those who were low on both avoidance and intrusion. Initial avoidance appears to be associated with subsequent increases in intrusive thoughts, perhaps reflecting failed efforts at suppressing unwanted thoughts.

Coping Responses

Evidence suggests that certain types of coping responses are associated with lower levels of psychological distress whereas others are related to higher distress. In general, patterns of coping that involve cognitive or behavioral avoidance of the disease and avoidance of negative thoughts and emotions associated with it have been found to be related to increased symptoms of anxiety and depression (e.g., Carver et al., 1993; Dunkel-Schetter, Feinstein, Taylor, & Falke, 1992; Epping-Jordan et al., 1997; Nelson et al., 1994; Stanton & Snider, 1993). For example, Stanton and Snider (1993) found that women who used more cognitive avoidance prior to their biopsy were more distressed after receiving a cancer diagnosis. Carver et al. (1993) found that initial optimism was related to the use of more acceptance, more humor, less denial, and less behavioral disengagement by women with breast cancer. The use of acceptance and humor as coping strategies was related to lower levels of psychological distress, whereas the use of denial and behavioral disengagement was related to higher distress. Thus, Carver et al. (1993) found that the effects of optimism on psychological adjustment to breast cancer were mediated by the types of coping efforts that are employed by women who are more optimistic.

Our conceptual model hypothesizes that interactive effects are important to consider, as different types of coping may vary in their effectiveness depending on contextual factors and cognitive appraisals. We have pursued this most directly in analyses of the interactions of coping and perceptions of control (Osowiecki & Compas, in press; Osowiecki, Compas, Epping-Jordan, & Krag, 1997). For example, we examined the interaction of perceptions of control over the progression of disease symptoms with each of four types of coping among newly diagnosed breast cancer patients (Osowiecki et al., 1997). The findings reveal both main effects for coping and interactions of coping with perceptions of control, with specific effects changing from diagnosis to 3- and 6-months post-diagnosis. These findings build on previous studies by our research group which have shown that distress is lower when problem-focused coping and control beliefs are matched (Compas et al., 1988; Forsythe & Compas, 1987). Coping processes need to be considered within the context of individuals' appraisals of the controllability of their cancer. Patients differ substantially in the degree to which they believe they can personally control or influence their disease, and patients who believe they have some degree of control will benefit more from the use of problem-focused coping than those who do not believe they have personal control.

An Integrated Model of Successful Adaptation to Cancer

These various aspects of adaptation to cancer are best considered in an overall model of the process of adjustment (Epping-Jordan et al., 1997). Demographic (age, education), disease (stage), and personality characteristics (optimism, monitoring) are exogenous variables. Cancer-related intrusive thoughts and avoidance of these thoughts are included next as proximal responses to diagnosis and treatment. The third step in the model includes the use of problem- and emotion-focused engagement and disengagement coping strategies. Finally, affective distress (symptoms of anxiety/depression) is included as the outcome. This model was tested 10 days post-diagnosis, 3 months post-diagnosis (during treatment), and at 6 months post-diagnosis (after completion of initial treatment for most patients). The general picture of the process of adjustment is relatively clear — cognitive and coping processes are crucial indicators of adjustment.

At the time of diagnosis, it appears that intrusive thoughts serve to mobilize more coping behavior, including both engagement and disengagement coping. The use of problem-focused engagement coping (problem-solving, cognitive restructuring) is related to less distress, whereas the emotion-focused disengagement (self-criticism, wishful thinking) is related to more distress. There is both a direct path from optimism to less distress and a mediated path through the use of less emotion-focused disengagement coping. The path model was tested again at 3 months, and the analyses suggest that during the most intensive period of treatment (chemotherapy, radiation therapy), coping processes exert little direct effect on changes in affective distress. Ongoing intrusive thoughts are the only factor related to increases in distress from diagnosis to 3 months. In the path model at 6 months post-diagnosis coping reemerges as an important predictor of changes in affective distress over this period. The effects of optimism are both direct and mediated by the use of emotion-focused engagement coping. Problem-focused engagement coping is related to decreases in distress, whereas both emotion-focused engagement and disengagement coping are related to increases in distress.

In addition to conducting group level analyses, we have considered individual differences in the patterns of temporal stability of psychological distress among breast cancer patients (Compas et al., 1997). These groups were distinguished in MANOVAs by their age, education, levels of pessimism, intrusive thoughts and avoidance, and the use of emotion-focused engagement and disengagement coping. Specifically, patients who were persistently high in distress were younger, had less education, were more pessimistic, had more intrusive thoughts and tried to avoid these thoughts, and used more emotion-focused engagement and disengagement coping. Linear discriminant function analyses indicate that classification of patients into these groups is best achieved by the use of optimism/pessimism and intrusive thoughts alone. Similar analyses were conducted to predict these patterns over the course of the first year after diagnosis, and initial intrusive thoughts were the only significant predictor of group differences over this period (Harding, Compas, & Oppedisano, 1997).

Findings from these studies of women with breast cancer have important implications for understanding competence in the face of stress and adversity. Competent functioning is reflected in both stable characteristics of cancer patients, and in the coping responses that they enact over the course of the stress of a cancer diagnosis and treatment. A realistic but optimistic cognitive style and the presence of a supportive social network serve as resources for enacting problem-focused engagement coping responses that are actively oriented toward the stressor. Engagement coping will be most effective, however, in dealing with those aspects of the stressor that are perceived as controllable, and at some points in the process (e.g., near diagnosis and after the completion of treatment) than others (e.g., during active treatment). Thus, competent functioning in the face of cancer is represented by a fluid process.

COPING AND COMPETENCE IN CHILD CANCER PATIENTS

Childhood cancer is diagnosed much less frequently than cancer in adults. However, a significant number of children, over 8,000, are diagnosed every year in the United States (American Cancer Society, 1996). It is a disease that is accompanied by frequent invasive medical procedures, such as injections, lumbar punctures, and venipunctures, as well as hospitalizations. As such, it is an acute stressor for most children. As recently as the 1970s, most children died from cancer; today, however, recent advances in treatments, in particular prophylactic treatment of the central nervous system (CNS) with intrathecal chemotherapy, CNS radiation therapy, or a combination of the two, have led to a dramatic improvement in the survival rates, with as many as 60% of diagnosed children surviving to adulthood. Given this change, it is predicted that by the year 2000, one in 1,000 adults, ages 20–29, will be a childhood cancer survivor (Meadows & Hobbie, 1986). While these survivors have a much greater life expectancy, they also face possible long-term effects of their treatment, as well as the increased risk of a second cancer. Pediatric cancer is therefore both an acute and a chronic stressor.

Psychological Adjustment to Pediatric Cancer

A large portion of this research on children's adjustment to cancer has focused on possible psychopathology in children, and has asked fewer questions about competence. Most of these studies of adjustment have shown that in spite of the challenges faced by children with cancer, these children are as well or better adjusted than other children (e.g., Brown, Kaslow, Hazzard, Madan-Swain, Sexson, Lambert, & Baldwin, 1992; Canning, Canning, & Boyce, 1992; Kazak, Barakat, Meeske, Christakis, Meadows, Casey, Penati, & Stuber, 1997; Noll, Ris, Davies, Bukowski, & Koontz, 1992; Olson, Boyle, Evans, & Zug, 1993; Varni, Katz, Colegrove, & Dolgin, 1996). For example, in their longitudinal study of adjustment to childhood cancer, Varni et al. (1996) found that internalizing and externalizing behavior problems were in the normal range at diagnosis and post-diagnosis. Other studies have shown similar results (Brown et al., 1992; Kazak, Christakis, Alderfer, & Coiro, 1994). Similarly, studies of depressive symptoms in children with cancer and survivors have shown very low symptom levels (Brown et al., 1992; Canning et al., 1992).

Most studies examining areas of competence have found that children with cancer perform at similar levels to normative samples. For example, Varni et al.'s (1996) study showed that children with cancer showed mean social competence T scores on the Child Behavior Checklist (CBCL; Achenbach, 1991) in the normative range at diagnosis and follow-up. Similarly, Noll et al. (1992) compared a group of children with cancer to a normative sample and found that children with cancer were rated by teachers more often as sociable and less often as aggressive–disruptive than a normative sample. Studies of self-concept in childhood cancer survivors have shown similar results. Anholt, Fritz, and Keener (1993) studied children who had survived cancer and a healthy sample, and found that both samples had global self-concepts in the normative range. In addition, they showed that cancer survivors had more positive feelings concerning their intellectual and school status, their behavior, and their happiness–satisfaction than did children in the comparison group. A recent study by Zeltzer et al. (1997) of adult survivors of ALL found that survivors experienced more symptoms of depression, tension, anger, and confusion compared with their sibling controls. However, the magnitude of the survivors' symptoms was not in the range of a clinical sample.

While childhood cancer survivors appear to be psychologically well adjusted, research has consistently shown declines in cognitive and neuropsychological functioning, as well as academic achievement, following CNS irradiation (Brown & Madan-Swain, 1993). Estimates in the literature suggest an average decrease in intellectual functioning of more than two-thirds of one standard deviation, particularly in the areas of visual-spatial and visual-motor skills, short-term memory, distractibility, attention, motor speed, and perception (Brown & Madan-Swain, 1993). The research on intrathecal chemotherapy is slightly less conclusive. Brown et al. (1996) failed to find any difference in cognitive functioning or achievement between a group which received intrathecal chemotherapy and a group which did not. However, the findings in the area of achievement approached significance and suggested a decrease of over one standard deviation in reading and spelling.

Despite the negative effects of treatment, the profile of cancer patients provided by the recent research is generally one of psychologically healthy, competent children and adolescents. Therefore, we can look at pediatric cancer patients and survivors as models of competence and examine what it is about these children and how they are coping with cancer that facilitates positive adjustment. Given what we know about adult adjustment to cancer, children's apparent competence following a diagnosis of cancer seems counter intuitive. How is it that these children can encounter such an acute, life-threatening stressor and can remain so well adjusted at diagnosis and many years later?

Denial, Repression, and Avoidant Coping

Some researchers (e.g., Canning et al., 1992; Phipps, Fairclough, & Mulhern, 1995) have suggested that the higher levels of adjustment reported in childhood cancer patients and survivors are a reflection of the use of defensive processes. Worchel et al. (1988) hypothesized that children with cancer use increased denial to handle problems, a strategy which minimizes their reported distress. Canning et al. (1992) found that in comparison to a healthy sample, a higher proportion of adolescent cancer patients were identified as repressors. They also showed that being a repressor was associated with lower levels of depression, and suggested that in adult populations, repression has been shown to have long term negative implications for health. Denial and repression can both be described as processes occurring outside of the individual's conscious awareness. In contrast, Phipps, Fairclough, and Mulhern (1995) looked at effortful or purposeful coping responses. In particular, they examined coping responses on a dimension that reflects one's tendency to approach or avoid information around a stressor and showed that children with cancer endorsed more blunting or avoidant coping than healthy children did.

The research on denial, repression, and avoidant coping raises questions about the apparent competence of childhood cancer survivors. In addition, these researchers argue that while repression and avoidant coping may be protective against psychological distress in the short term, they can become harmful in a more chronic stressful situation, leading to impaired adherence to treatment recommendations and increased distress. Thus, this research suggests that there may be long term negative effects of a defensive style. Most current research does not support this hypothesis, however, with long term (up to 10 years post treatment) positive psychological outcomes in childhood cancer survivors (Kazak et al., 1997; Kupst et al., 1995).

An alternative approach to the research on denial, repression and avoidant coping is to look at the behaviors which accompany these coping styles. Distraction, frequently paired with avoidant behavior, is a coping strategy which involves not only avoidance of the stressor, but also the active shifting of one's attention to more positive or neutral targets. Unlike avoidance, distraction has been shown to be effective and is associated with positive adjustment. In particular, distraction, and other efforts to modify one's own behavior to fit the situation, have been shown to be more efficacious with low controllability stressors, such as cancer and associated treatments, than primary control efforts, or efforts to modify the objective situation (Weisz, McCabe, & Dennig, 1994). Research has also shown that distraction is an adaptive coping style frequently used by children to effectively manage painful medical procedures (Manne, Bakeman, Jacobsen, & Redd, 1993) and chemotherapy-induced nausea (Tyc, Mulhern, Jayawardene, & Fairclough, 1995). While distraction is frequently associated with avoidant behavior, it is sometimes not measured or is confused with avoidant behavior. The failure to measure distraction and the confusion between avoidance and distraction may have led to erroneous conclusions regarding the coping strategies used by children with cancer. The use of distraction, as opposed to denial, repression, and avoidance, may represent effective coping and may account for the positive adjustment in pediatric cancer patients.

Parental Coping

Another possible explanation for the positive adjustment of children with cancer is that parents help to manage their children's stress and serve as protective factors. One way to examine this hypothesis is to look at parents' coping efforts and the role they play in the adjustment of the children. Sanger, Copeland, and Davidson (1991) examined the relationship between parental coping and children's adjustment and found that parents whose children were rated as better adjusted reported greater efforts to maintain family integration and an optimistic definition of the illness than parents of children less well adjusted. Thus, Sanger et al. suggest that parents who show hopeful attitudes and facilitate positive relationships within the family may model important coping strategies and provide reassurance for their children. Kupst et al. (1995) also examined parental coping in their prospective study of family coping. At the 10 year assessment, the most significant predictors of survivors' coping and adjustment were the coping and adjustment of the mother. Interestingly, the child's coping behavior and adjustment was also a predictor of the mother's adjustment.

Similarly, Varni et al. (1996) showed that the family dimensions of cohesion and expressiveness predicted the psychological and social adjustment of children over a 9-month period. Thus, supportive family environments which encourage the open communication of thoughts and feelings facilitate the adjustment process to cancer.

Cognitive Development

A third explanation for the positive adjustment of children with cancer focusses on the role of cognitive development in the ability of children to understand cancer as a stressor. Kazak et al. (1997) found that while parents of childhood leukemia survivors experienced significant post traumatic stress symptoms (10.2% of mothers and 9.8% of fathers), the survivors did not experience any more symptoms than a comparison sample. Kazak et al. noted that the median age at diagnosis was 3–4 years in their sample, and suggested that many survivors would have limited memories of the event, and thus were not at risk for PTSD symptoms. In addition, Kazak et al. (1997) found that the child's current age and age at diagnosis were not related to PTSD symptoms. Kazak et al. (1997) argued that the limited ability of preschool and school age children to process the life threat of cancer might serve to protect them from the trauma. In contrast, their parents, who were fully aware of the life threat implied by a cancer diagnosis, were more psychologically distressed by the event.

In a related vein, research in the area of children's understanding of illness has focused on the role of children's level of cognitive development. Studies have shown that children comprehend more about illness and become less concrete in their definition of illness with age (Campbell, 1975). Experience has also been shown to play an important role in the understanding of illness. Crisp, Ungerer, and Goodnow (1996) examined the role of experience in two groups of children, one with cystic fibrosis and one with cancer. Their results point to both age and experience as having a facilitating effect on understanding for children 7 to 10 years of age, and children who are diagnosed at 3 to 4 years with little illness experience are likely to have a limited understanding of the meaning of their illness. Thus, children may have problems processing not only the life threat aspect of their illness, suggested by Kazak et al. (1997), but also the meaning of the illness itself. Such cognitive limitations may play a significant role in facilitating positive adjustment.

Children's Coping Efforts

The theories for positive adjustment discussed above all focus either on external sources (e.g., parental coping) or on processes outside of the child's awareness (e.g., defensive processes, cognitive development). An alternative conceptualization of the positive adjustment of these children centers on the cognitive and behavioral efforts they are making to manage the stress of cancer. For example, Brown et al. (1992) found that the children with cancer used strategies of problem solving, wishful thinking and emotional regulation most often. In addition, their analyses showed that with age, children used more distraction, cognitive restructuring, emotional regulation, social support, and resignation. Worchel, Copeland, and Barker (1987) examined coping by asking children about their strategies for gaining control and found that higher behavioral control was associated with poorer adjustment. Worchel et al. (1987) suggest that there are a number of different strategies on the behavioral control scale, and argue that the high scorers are trying many different methods and coping ineffectively. This is consistent with the literature suggesting that the use of primary control coping strategies with low-controllability stressors leads to poorer adjustment (Weisz et al., 1994).

Another source of information regarding children's coping strategies lies in the literature regarding painful medical procedures. For example, Blount et al. (1989) examined the behaviors of pediatric oncology patients undergoing bone marrow aspirations and lumbar punctures, as well as their parents and the medical staff. The procedures were coded and children were grouped into high and low coping groups. Results showed that adults made significantly more coping promoting statements in the high coping group, and that the high coping children were more likely to engage in distracting conversation and deep breathing following adult coping promoting statements. These results suggest that adults play an important role in facilitating the coping process for children with cancer.

The literature on interventions for painful procedures gives us some insight into which strategies, in general, are helpful for children. In a review of cognitive–behavioral interventions for children's distress during bone marrow aspirations and lumbar punctures, Ellis and Spanos (1994) concluded that children can gain some control over their discomfort by controlling and redirecting their thoughts. Powers, Blount, Bachanas, Cotter, and Swan (1993) delivered an intervention to four preschool leukemia patients in a multiple baseline design. In the intervention, children were taught to engage in distraction activities prior to the procedures, deep breathing, and to use counting strategies during the procedures. The parents were taught to engage in the distraction and to coach on the breathing and counting. After intervention, relative to baseline, parents were much more engaged in promoting coping behaviors, children were using more coping strategies, and child behavioral distress was lowered.

Research in the area of interventions for medical procedures offers insight into the importance of distraction and other coping techniques which help the child to disengage from the situation. With a low-controllability stressor, like cancer, it appears that secondary control coping techniques, which are oriented towards maximizing one's fit to the situation, are more successful. In their review of coping in the medical setting, Rudolph, Dennig, and Weisz (1995) concluded that when coping with a medical stressor, older children are more likely to use secondary control or cognitively based strategies than younger children. They suggested that younger children may have more problems generating cognitively based strategies, may be less effective with them, and may require more guidance in the process. Since most childhood cancer diagnoses are at this early age further underscores the importance of parental coping in facilitating the child's adjustment.

COPING AND COMPETENCE IN FAMILIES OF CANCER PATIENTS

Although the majority of research on the psychosocial aspects of cancer has focused on the adjustment of adults and children as patients, there is wide recognition that cancer affects more than just patients. Cancer can have a profound effect on the lives of family members and others who are close to patients. Research on the ways that families cope with cancer has come from two sources — studies of families in which a mom or dad has cancer, and studies of families in which a child is diagnosed with cancer. Although they are reflected in separate research literatures, studies of parental and child cancer both share an emphasis on the well being and the coping efforts of those who are closely tied to the patient.

A growing body of evidence now suggests that parental cancer is a significant stressor for children and adolescents (e.g., Compas et al., 1994; Compas, Worsham, & Howell, 1996; Grant & Compas, 1995; Lewis, Hammond, & Woods, 1993; Welch, Wadsworth, & Compas, 1996). There are substantial individual differences in adjustment to parental cancer as a function of the age and gender of the child, whether the mother or the father has cancer, and the source of information about the child's adjustment. In our own research, we assessed anxiety/depression symptoms and stress response syndrome symptoms (intrusive thoughts and avoidance) as indices of psychological distress in adult cancer patients, spouses, and their children (ages 6 to 30 years old) near the time of patients' diagnoses to identify family members most at risk for psychological maladjustment. Patients' and family members' symptoms of distress were generally unrelated to objective characteristics of the disease but were related to appraisals of the seriousness and stressfulness of the cancer (Compas et al., 1994). However, both stress response and anxiety/depression symptoms differed in children as a function of their age, and the interaction of sex of child and sex of patient. Specifically, adolescent girls whose mothers had cancer were the most significantly distressed group of children. Furthermore, the distress experienced by adolescents in these families was evident in their self-reports of anxiety and depression, but not in their parents' reports on their children's adjustment (Welch et al., 1996).

These findings raised two complementary questions — why are adolescent girls whose mothers have cancer so much more distressed than their siblings, and why are younger children and adolescent boys not significantly distressed? In an analysis of possible mechanisms to account for the effects of maternal cancer on adolescent daughters, we examined coping responses used by adolescent girls and the types of ongoing stresses and strains within the families that were the result of the parents' diagnosis and treatment. Specifically, we tested the role of ruminative coping (Nolen-Hoeksema, 1991) and increased responsibilities for taking care of others in the family as risk factors for adolescent girls' heightened symptoms of anxiety and depression (Grant & Compas, 1995). Analyses revealed that increased caretaking responsibilities experienced by girls whose mothers were ill accounted for their higher levels of psychological distress (Grant & Compas, 1995). Thus, it appears that disruptions in the roles and responsibilities were particularly stressful for adolescent girls, and that these girls paid a price for these disruptions as evidenced in their increased levels of affective distress.

SUMMARY AND INTEGRATION

Research on the process of psychosocial adaptation to cancer provides a number of important insights into the nature of competent functioning in the context of extreme stress and adversity. Several themes have emerged from studies of children, adults, and families that offer a perspective on competence and coping with a serious illness across the lifespan.

First, the diagnosis and treatment of cancer presents individuals and their families with a series of multiple stressors that extend over weeks and months, rather than a discrete stressful event that is limited to a specific point in time. Although the diagnosis itself may represent a singularly traumatic event, the stress that is entailed in treatment, recovery, and long-term living with uncertainty appears to be more adverse than the original trauma of the diagnosis. Furthermore, the number and severity of the stressful events associated with cancer are significant — positive adaptation by patients and their families cannot be attributed to an absence of threats and challenges associated with cancer.

Second, in spite of the significance of cancer as a stressor, most individuals and their families achieve positive psychological outcomes as reflected in measures of emotional distress, quality of life, and functioning in social, academic, and work domains. The general picture is one of highly competent functioning in the face of adversity. Positive outcomes are evident for the majority of adult cancer patients, children with cancer, and family members of both child and adult cancer patients.

Third, the adjustment of adults reflects higher levels of emotional distress near the time of their diagnosis and treatment, but these initial levels of distress subside over time. Thus, effective coping must be viewed as a process that changes with time and the changing demands of the stressor. Positive outcomes are achieved by the majority of patients, even if initial levels of distress are high.

Fourth, in spite of the overall positive picture of psychological adjustment, there are significant numbers of patients and their family members who experience both short- and long-term problems in adjustment. These problems are reflected in heightened symptoms of depression and anxiety, decreased quality of life, and disrupted functioning in life domains. Variations in the quality of adjustment are associated with development across the lifespan. That is, children with cancer appear to experience relatively fewer psychological difficulties than adults with cancer, and among adults, older adults experience fewer problems than young adults. Among family members, parents of children with cancer, and children, especially adolescents, whose parents have cancer also experience significant adjustment problems.

Fifth, these studies highlight the importance of studying both adaptive as well as maladaptive functioning in order to derive a complete picture of the nature of adjustment to extreme stress. Research on stress has typically emphasized pathology and maladjustment, and has failed to include measures of competent functioning. This emphasis on negative outcomes would result in a distorted picture of the process adaptation to stress, as indicated by the findings of high levels of competence among children with cancer, children whose parents have cancer, and the majority of adults who are diagnosed with the disease.

Finally, and most important to our understanding of competence and coping, variation between successful and unsuccessful adaptation is associated with specific personal and social resources, and with the use of certain coping strategies and responses. Research with children and adults has emphasized the importance of a realistic but positive attitude and the presence of supportive family relationships as resources for effective coping with cancer. Coping represents resources and competence that have been put into action in the face of significant stress. Effective coping changes as the demands of a stressor change, and as a function of the developmental level of the individual. Research on coping with cancer provides an informative model for research on coping with a variety of other stressors, and the clues to competent functioning in a wide range of contexts.

REFERENCES

Achenbach, T. M. (1991). *Manual for the child behavior checklist/4-18 and 1991 profile*. Burlington, University of Vermont Department of Psychiatry.

American Cancer Society. (1996). *Cancer facts and figures — 1996*. Atlanta: Author.

American Psychiatric Association (1994). *Diagnostic and statistical manual of mental disorders* (4th ed.). Washington, DC.: Author.

Anholt, U. V., Fritz, G. K., & Keener, M. (1993). Self-concept in survivors of childhood and adolescent cancer. *Journal of Psychosocial Oncology, 11*, 1–16.

Banks, E. (1990). Concepts of health and sickness of preschool and school-aged children. *Children's Health Care, 19*, 43–48.

Baum, A., Cohen, L., & Hill, M. (1993). Control and intrusive memories as possible determinants of chronic stress. *Psychosomatic Medicine, 55*, 274–286.

Bloom, J. R. (1986). Social support and adjustment to breast cancer. In B. L. Andersen (Ed.), *Women with cancer: Psychological perspectives* (pp. 204–229). New York: Springer.

Bloom, J. R., & Spiegel, D. (1984). The relationship of two dimensions of social support to the psychological well-being and social functioning of women with advanced breast cancer. *Social Science and Medicine, 19*, 831–837.

Blount, R. L., Landolf-Fritsche, B., Powers, S. W., & Sturges, J. W. (1991). Differences between high and low coping children and between parent and staff behaviors during painful medical procedures. *Journal of Pediatric Psychology, 16*, 795–809.

Blount, R. L., Corbin, S. M., Sturges, J. W., Wolfe, V. V., Prater, J. M., & James, L. D. (1989). The relationship between adults' behavior and child coping and distress during BMA/LP procedures: A sequential analysis. *Behavior Therapy, 20*, 585–601.

Brown, R. T., Kaslow, N. J., Hazzard, A. P., Madan-Swain, A., Sexson, S. B., Lambert R., & Baldwin, K. (1992). Psychiatric and family functioning in children with leukemia and their parents. *Journal of the American Academy of Child and Adolescent Psychiatry, 31*, 495–502.

Brown, R. T. & Madan-Swain, A. (1993). Cognitive, neuropsychological and academic sequelae in children with leukemia. *Journal of Learning Disabilities, 26*, 74–90.

Brown, R. T., Sawyer, M. B., Antoniou, G., Toogood, I., Rice, M., Thompson, N., & Madan-Swain, A. (1996). *Developmental and Behavioral Pediatrics, 17*, 392–398.

Campbell, J. D. (1975). Illness is a point of view: The development of children's concepts of illness. *Child Development, 46*, 92–100.

Canning, E. H., Canning, R. D., & Boyce, W. T. (1992). Depressive symptoms and adaptive style in children with cancer. *Journal of the American Academy of Child and Adolescent Psychiatry, 31*, 1120–1124.

Carver, C. S., Pozo, C., Harris, S. D., Noriega, V., Scheier, M. F., Robinson, D. S., Ketcham, A. S., Moffat, F. L., & Clark, K. C. (1993). How coping mediates the effects of optimism on distress: A study of women with early stage breast cancer. *Journal of Personality and Social Psychology, 65*, 375–391.

Carver, C. S., Pozo-Kaderman, C., Harris, S. D., Noriega, V., Scheier, M. F., Robinson, D. S., Ketcham, A. S., Moffat, F. L., & Clark, K. C. (1994). Optimism versus pessimism predicts the quality of women's adjustment to early stage breast cancer. *Cancer, 73*, 1213–1220.

Compas, B. E. (1993). Promoting positive mental health during adolescence. In S. G. Millstein, A. C. Petersen, & E. O. Nightingale (Eds.), *Promoting the health of adolescents: New directions for the twenty-first century.* New York: Oxford University Press.

Compas, B. E., Connor, J., Harding, A., Saltzman, H., & Wadsworth, M. (in press). Getting specific about coping: Effortful and involuntary responses to stress in development. In M. Lewis & D. Ramsey (Eds.), *Stress and soothing.*

Compas, B. E., Conner, J., Osowiecki, D., & Welch, A. (1997). Effortful and involuntary responses to stress. In B. Gottlieb (Ed.), *Coping with chronic stress.* New York: Plenum.

Compas, B. E., Malcarne, V., & Fondacaro, K. (1988). Coping with stress in older children and young adolescents. *Journal of Consulting and Clinical Psychology, 56*, 405–411.

Compas, B. E., Oppedisano, G., Primo, K., Epping-Jordan, J. E., Gerhardt, C. A., Osowiecki, D. M., & Krag, D. N. (1997). Psychological adjustment to breast cancer: Markers of subgroup differences in emotional distress. Manuscript submitted for publication.

Compas, B. E., Worsham, N., Epping-Jordan, J. E., Howell, D. C., Grant, K. E., Mireault, G., & Malcarne, V. (1994). When mom or dad has cancer: Markers of psychological distress in cancer patients, spouses, and children. *Health Psychology, 13*, 507–515.

Compas, B. E., Worsham, N., Ey, S., & Howell, D. C. (1996). When mom or dad has cancer: II. Coping, cognitive appraisals, and psychological distress in children of cancer patients. *Health Psychology, 14.*

Crisp, J., Ungerer, J. A., & Goodnow, J. J. (1996). The impact of experience on children's understanding of illness. *Journal of Pediatric Psychology, 21*, 57–72.

Dakof, G. A., & Taylor, S. E. (1990). Victims' perceptions of social support: What is helpful from whom? *Journal of Personality and Social Psychology, 58*, 80–89.

Derogatis, L. R., Morrow, G. R., Fetting, J., Penman, D., Piasetsky, S., Schmale, A. M., Henrichs, M., & Carnicke, C. L. M. (1983). The prevalence of psychiatric disorders among cancer patients. *Journal of the American Medical Association, 249*, 751–757.

Dunkel-Schetter, C., Feinstein, L. G., Taylor, S. E., & Falke, R. L. (1992). Patterns of coping with cancer. *Health Psychology, 11*(2), 79–87.

Ellis, J. A., & Spanos, N. P. (1994). Cognitive–behavioral interventions for children's distress during bone marrow aspirations and lumbar punctures: A critical review. *Journal of Pain and Symptom Management, 9*, 96–108.

Epping-Jordan, J. E., Compas, B. E., & Howell, D. C. (1994). Psychological symptoms, avoidance, and intrusive thoughts as predictors of cancer progression. *Health Psychology, 13*, 539–547.

Epping-Jordan, J. E., Compas, B. E., Osowiecki, D., Oppedisano, G., Gerhardt, C., & Krag, D. (1997). Predictors of psychological adjustment to breast cancer: Disease, demographic, personality, and coping factors. Manuscript submitted for publication.

Fallowfield, L. J., Hall, A., Maguire, G. P., & Baum, M. (1990). Psychological outcomes of different treatment policies in women with early breast cancer outside a clinical trial. *British Medical Journal, 301*, 575–580.

Forsythe, C. J. & Compas, B. E. (1987). Interaction of cognitive appraisals of stressful events and coping: Testing the goodness of fit hypothesis. *Cognitive Therapy and Research, 11*, 473–485.

Funch, D. P., & Marshall, J. (1983). The role of stress, social support, and age in survival from breast cancer. *Journal of Psychosomatic Research, 27*, 77–83.

Ganz, P. A., Hirji, K., Sim, M., Schag, C. A. C., Fred, C., & Polinsky, M. L. (1993). Predicting psychosocial risk in patients with breast cancer. *Medical Care, 31*, 419–431.

Glanz, K., & Lerman, C. (1992). Psychosocial impact of breast cancer: A critical review. *Annuals of Behavioral Medicine, 14*, 204–212.

Grant, K. E., & Compas, B. E. (1995). Stress anxious/depressed symptoms among adolescents: Searching for mechanisms of risk. *Journal of Consulting and Clinical Psychology, 63*, 1015–1021.

Harding, A., Compas, B. E., & Oppedisano, G. (1997). Individual differences in psychological adjustment to breast cancer over two years. Paper presented at the annual meeting of the Society of Behavioral Medicine, San Francisco, CA.

Hilton, B. A. (1989). The relationship of uncertainty, control, commitment, and threat of recurrence to coping strategies used by women diagnosed with breast cancer. *Journal of Behavioral Medicine, 12*, 39–54.

Horowitz, M. J. (1982). Stress response syndromes and their treatment. In L. Goldberger & S. Breznitz (Eds.), *Handbook of stress* (pp. 711–732). New York: Free Press.

Kazak, A. E., Barakat, L. P., Meeske, K., Christakis, D., Meadows, A. T., Casey, R., Penati, B., & Stuber, M. L. (1997). Posttraumatic stress, family functioning, and social support in survivors of childhood leukemia and their mothers and fathers. *Journal of Consulting and Clinical Psychology, 65*, 120–129.

Kazak, A. E., Christakis, D., Alderfer, M., & Coiro, M. J. (1994). Young adolescent cancer survivors and their parents: Adjustment, learning problems, and gender. *Journal of Family Psychology, 8*, 74–84.

Kupst, M. J., Natta, M. B., Richardson, C. C., Schulman, J. L., Lavigne, J. V., & Das, L. (1995). Family coping with pediatric leukemia: Ten years after treatment. *Journal of Pediatric Psychology, 20*, 601–617.

Lazarus, R. S., & Folkman, S. (1984). *Stress, appraisal, and coping.* New York: Springer.

Levy, S. M., Herberman, R. B., Whiteside, T., Sanzo, K., Lee, J., & Kirkwood, J. (1990). Perceived social support and tumor estrogen/progesterone receptor status as predictors of natural killer cell activity in breast cancer patients. *Psychosomatic Medicine, 52*, 73–85.

Lewis, F. M., Hammond, M. A., & Woods, N. F. (1993). The family's functioning with newly diagnosed breast cancer in the mother: The development of an exploratory model. *Journal of Behavioral Medicine, 16*, 351–370.

Manne, S. L., Bakeman, R., Jacobsen, P., & Redd, W. H. (1993). Children's coping during invasive medical procedures. *Behavior Therapy, 24*, 143–158.

Maston, A. S., Coatsworth, J. D., Neeman, J., Gest, S. D., Tellegen, A., & Garmezy, N. (1995). The structure and coherence of competence from childhood through adolescence. *Child Development, 66*, 1635–1659.

Masten, A. S., Hubbard, J. J., Gest, S. D., Tellegen, A., Garmezy, N., & Ramirez, M. (1997). Adaptation in the context of adversity: Pathways to resilience and maladaptation from childhood to late adolescence. Manuscript submitted for publication.

Meadows, A. T., & Hobbie, W. L. (1986). The medical consequences of cure. *Cancer, 58*, 524–528.

Morris, T., Greer, H., & White, P. (1977). Psychological and social adjustment to mastectomy: A two-year follow-up. *Cancer, 77*, 2381–2387.

Nelles, W. B., McCaffrey, R. J., Blanchard, C. G., & Ruckdelschel, J. C. (1991). Social supports and breast cancer: A review. *Journal of Psychosocial Oncology, 9*, 21–34.

Nelson, D. V., Friedman, L. C., Baer, P. E., Lane, M., & Smith, F. E. (1994). Subtypes of psychosocial adjustment to breast cancer. *Journal of Behavioral Medicine, 17*, 127–141.

Noll, R. B., Ris, M. D., Davies, W. H., Bukowski, W. M., & Koontz, K. (1992). Social interactions between children with cancer or sickle cell disease and their peers: Teacher ratings. *Developmental and Behavioral Pediatrics, 13*, 187–193.

Nolen-Hoeksema, S. (1991). Responses to depression and their effects on the duration of depressive episodes. *Journal of Abnormal Psychology, 100*, 569–582.

Olson, A. L., Boyle, W. E., Evans, M. W., & Zug, L. A. (1993). Overall function in rural childhood cancer survivors. *Clinical Pediatrics, 32*, 334–342.

Osowiecki, D., & Compas, B. E. (in press). Coping and control beliefs in adjustment to cancer. *Cognitive Therapy and Research.*

Osowiecki, D., Compas, B. E., Epping-Jordan, J. E., & Krag, D. (1997). A prospective analysis of coping and control beliefs in psychosocial adjustment to breast cancer. Unpublished manuscript, University of Vermont.

Phipps, S., Fairclough, D., & Mulhern, R. K. (1995). Avoidant coping in children with cancer. *Journal of Pediatric Psychology, 20*, 217–232

Powers, S. W., Blount, R. L., Bachanas, P. J., Cotter, M. W., & Swan, S. C. (1993). Helping preschool leukemia patients and their parents cope during injections. *Journal of Pediatric Psychology, 18*, 681–695.

Primo, K., Compas, B. E., Oppedisano, G., Epping-Jordan, J. E., & Krag, D. (1997). Patterns of individual differences in intrusive thoughts and avoidance in adjustment to breast cancer. Manuscript submitted for publication.

Rudolph, K. D., Dennig, M. D., & Weisz, J. R. (1995). Determinants and consequences of children's coping in the medical setting: Conceptualization, review, and critique. *Psychological Bulletin, 118*, 328–357.

Sanger, M. S., Copeland, D. R., & Davidson, E. R. (1991). Psychosocial adjustment among pediatric cancer patients: A multidimensional assessment. *Journal of Pediatric Psychology, 16*, 463–474.

Schag, C. A. C., Ganz, P. A., Polinsky, M. L., Fred, C., Hirji, K., & Peteresen, L. (1993). Characteristics of women at risk for psychological distress in the year after breast cancer. *Journal of Clinical Oncology, 11*, 783–793.

Stanton, A. L., & Snider, P. R. (1993). Coping with a breast cancer diagnosis: A prospective study. *Health Psychology, 12*, 16–23.

Stoll, M. F., Oppedisano, G., Epping-Jordan, J. E., Compas, B. E., & Krag, D. N. (1997). Adjustment to breast cancer: Age-related differences in coping and emotional distress. Unpublished manuscript, University of Vermont.

Tyc, V. L., Mulhern, R. K., Jayawardene, D., & Fairclough, D. (1995). Chemotherapy-induced nausea and emesis in pediatric cancer patients: An analysis of coping strategies. *Journal of Pain and Symptom Management, 10*, 338–347.

Varni, J. W., Katz, E. R., Colegrove, R., & Dolgin, M. (1996). Family functioning predictors of adjustment in children with newly diagnosed cancer: A prospective analysis. *Journal of Child Psychology & Psychiatry & Allied Disciplines, 37,* 321–328.

Vinokur, A. D., Threatt, B. A., Vinokur-Kaplan, D., & Satariano, W. A. (1990). The process of recovery from breast cancer for younger and older patients: Changes during the first year. *Cancer, 65,* 1242–1254.

Wegner, D. M. (1994). Ironic processes of neutral control. *Psychological Review, 101,* 134–152.

Weisz, J. R., McCabe, M. A., & Dennig, M. D. (1994). Primary and secondary control among children undergoing medical procedures: Adjustment as a function of coping style. *Journal of Consulting and Clinical Psychology, 62,* 324–332.

Welch, A. S., Wadsworth, M. E., & Compas, B. E. (1996). Adjustment of children and adolescents to parental cancer. *Cancer, 77,* 1409–1418.

Worchel, F. F., Nolan, B. F., Wilson, V. L., Purser, J. S., Copeland, D. R., & Pfefferbaum, B. (1988). Assessment of depression in children with cancer. *Journal of Pediatric Psychology, 13,* 101–112.

Worchel, F. F., Copeland, D. R., Barker, D. G. (1987). Control-related coping strategies in pediatric oncology patients. *Journal of Pediatric Psychology, 12,* 25–38.

Zeltzer, L. K., Chen, E., Weiss, R., Guo, M. D., Robison, L. L., Meadows, A. T., Mills, J. L., Nicholson, H. S., & Byrne, J. (1997). Comparison of psychologic outcome in adult survivors of childhood acute lymphoblastic leukemia versus sibling controls: A cooperative Children's Cancer Group and National Institutes of Health Study. *Journal of Clinical Oncology, 15,* 547–556.

2

EMOTIONAL COMPETENCE AND DEVELOPMENT

Michael Lewis

Institute for the Study of Child Development
Robert Wood Johnson Medical School

In this chapter, the question of emotional development is addressed, keeping in mind that emotional behaviour, often considered the content of social interactions, is very much related to what we think of as a child's competence. Beside the obvious, namely that emotional dysfunction is a major part of psychopathology, emotional competence is very much related to motivation (Izard, 1977), for example, a child's emotional life is likely to influence such areas as school performance. Following a discussion of competence, I then focus on emotional development (Lewis, 1992; 1993a; 1993b).

WHAT IS COMPETENCE?

When talking about emotional competence our first task is to try and define competence. I am a little troubled by the term "competence" because like "intelligence," while it has common meaning, it is not clear what it means or how to measure it. Is it something like mental health, doing well in school, or succeeding in relationships? It could be all of these or any one. I prefer to think about competence from an evolutionary perspective and consider the term "adaptation," a similar idea to "competence." When we consider competence as adaptation, especially from an evolutionary point of view, we notice that adaptation refers to the here and now. Evolutionary theory is not predictive (Lewis, 1997). What is successful adaptation at the present time says nothing about successful adaptation later. Certainly this may apply to competence as well. This, of course, depends on whether we think of competence as a trait. If competence is a trait which some children have more of than others, then we are defining competence as a characteristic which a child possesses across situations. On the other hand, competence could be situationally determined. For example, there was a student who was excellent in mathematics, but could not spell. If we defined competency by spelling ability, he would be incompetent, but if we defined it by mathematics, he would excel. Perhaps the example of savant best captures the problem with defining competence. Savants can excel in one aspect of their intellectual function-

Improving Competence across the Lifespan, edited by Pushkar *et al.*
Plenum Press, New York, 1998.

ing, but be incompetent in most others. Such examples suggest that different competencies may be quite independent. While competence may or may not be a single trait, incompetence on the other hand is usually considered a trait.

We tend to think of competence in the very young child as deriving from early social relationships (Bowlby, 1969). Thus, the idea of attachment as a source of competence has been suggested; secure children are competent, insecure children are not. Such asymmetries as incompetence being a trait or individual characteristic while competence is more situationally specific raises the question as to the meaning of competence as well as to ask where does it, or they, come from and in what contexts are we to measure them? For me, attachment is not the place to look for competence. Emotional life as a general competence would seem to be a better choice because it seems that the content of all social exchange is emotion.

In this chapter competence is thought about as the acquisition and regulation of emotion. To begin to understand this competence we need a theory about emotional development. Any model of emotional development is going to need to consider species specific maturational processes. No matter how we try to teach a goldfish to talk, there is nothing in the environment that can produce this skill. Thus, there are likely to be species specific maturational processes which contribute to emotional development. Second, we need to consider the idea of individual differences. In our species, it is likely, at least as a starting point, to assume individuals form a normal distribution around particular species specific behaviors. Third, we need to consider socialization factors as they impact on emotional life. Finally, it must be understood that emotional, social, and cognitive skills are all aspects of a particular child and that they are related in complex ways.

EMOTIONAL DEVELOPMENT

The theory of emotional development that I have been working on has as its theme something from the Old Testament, out of Genesis. As you recall in the Garden of Eden, Adam and Eve were instructed not to eat of the tree of knowledge. Eve was tempted by the serpent; they ate of the tree of knowledge and they discovered that they were naked. They hid from one another and from God because they were naked. To reduce the story to its bare facts, we can see a developmental sequence: Adam and Eve are created with interest and curiosity. Eating the fruit, they gain knowledge, and from gaining knowledge, they gain the new emotion of shame. To formalize this story, we can say that emotions lead to cognitions, which in turn lead to new emotions. This is at the heart of the theory that follows.

Figure 1 presents a scheme useful in explaining emotional development. First notice what has been referred to as the primary emotions (Darwin, 1969/1872; Izard, 1977). They are primary because they appear first and include the emotions of joy, fear, anger, pride, disgust, and sadness. Some theorists suggest even more basic emotions (Tomkins, 1963). These emotions and how they are responded to affect the cognitive capacity that is called objective self-awareness or what Lewis (1995a) prefers to call consciousness; that is, the child's representation of itself. It is most likely that this cognitive capacity is the result of a maturation of certain parts of the cerebral cortex (Lewis, 1995a). This capacity emerges between 15 and 24 months of age.

The study of this acquisition has been considered by observing three very different sets of behaviors. First, we examine the infant's self-recognition, arguing that when the child looks at its reflection in the mirror s/he says in effect, "That's me in the mirror" (Lewis & Brooks-Gunn, 1979). In addition to mirror recognition, Lewis and Ramsay (in

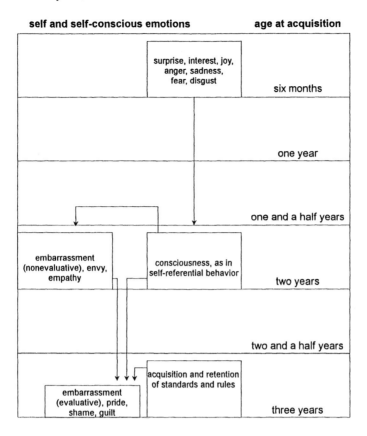

self and self-conscious emotions   **age at acquisition**

Figure 1. A model of emotional development. [Figure taken from Lewis, M. (1995). Self-conscious emotions. *American Scientist*, *83*, 68–78.]

press) have collected data on personal pronoun usage and pretend play. It is well known that at the same time that self-recognition emerges, pronouns such as "me" and "mine" emerge (Lewis & Michalson, 1983). Pretend play is another behaviour where consciousness is necessary. Pretend play is remarkable since it presumes that the child knows that what it plays with is not what the object really is; for example, if John pretends a pencil is an airplane, and John does not know it is a pencil, then John is either not very smart or he is having an hallucination. Pretend play, like other ways of studying theories of mind, involves a negation; "I know it is not an airplane, but I will pretend it is." It is the negation which implies consciousness.

The work we have done to date on consciousness is summarized in Lewis (1995d). Briefly reviewed, what we have found is presented below. Three-month-old children show a great deal of pleasure looking at their image, but this is not self-recognition. Infants, like other children, as well as animals and adult humans, find enjoyment in seeing pictures of infants. When the proper controls are initiated, we find that 3-month-old infant's responses to themselves in the mirror are similar to their responses to any infant (Lewis & Brooks-Gunn, 1979). By 8-months when children look in the mirror they often try to locate the child whose image they see. Some children crawl behind the mirror to find the child they see in the mirror. Interestingly, domestic animals, like cats and dogs, if they look in the mirror, often will go around the mirror to find the other animal. Between 15- and 24-

months, behaviors emerge which for us reflect the recognition of self in the mirror. Children of this age range, while looking in the mirror, show self-directed behaviour; that is children's noses are marked with rouge prior to their looking in the mirror and at this age, but not before, these children use the mirror to touch their noses. They also use personal pronouns such as "mine" or "me" (Lewis & Ramsay, in press).

Hundreds of infants have been tested using this procedure and these age changes have been found. Children need a mental age of 18 months in order to recognize themselves. Children with less than 18-months mental age do not recognize themselves in the mirror. Thus, for example, children with mental retardation who do not have a mental age of 18-months, even though they have a chronological age over 18-months, will not touch their noses. This evidence among others, leads to the belief that self-recognition in mirrors is due to maturational processes. This capacity emerges with a narrow window between 15- and 24-months. We believe it may be related to the myelination of the frontal lobe, one of the last areas to be myelinated in infants. This is consistent with the clinical literature which relates damage in the frontal lobes to self-regulation (Prigatano, 1991). Somewhere in the second half of the second year of life, consciousness emerges.

Individual differences in recognition exist within this six-month window of acquisition. Self-recognition appears to be related to maturational processes while individual differences in the rate of acquisition appears to be related to the temperament of the child. The difficult temperament child, either measuring temperament by parental report or measuring it by actually looking at children's stress responses to inoculation, is more likely to have earlier self-recognition than is the child without a difficult temperament (Lewis & Ramsay, 1997). Thus, individual differences in the timing of emergence of this capacity are related both to maturation and to individual differences in the ability to self-regulate, a common difficulty found in children with difficult temperament.

As can be seen in Figure 1, once the onset of self-recognition occurs, reflecting the underlying process of consciousness, new emotional capacities also emerge. In a series of studies (Lewis, Sullivan, Stanger, & Weiss, 1989), we have been able to show that embarrassment emerges only when the child has self-recognition capacity. Children do not show embarrassment prior to the emergence of self-recognition. The reason that embarrassment is related to self-recognition is that one can only be embarrassed if one has the idea of self, since embarrassment occurs when the child (or adult, for that matter) becomes the object of other people's attention (see Darwin, 1969/1872). Individual differences in embarrassment also appear to be related to temperament. DiBiase and Lewis (1997) have shown that infants show more embarrassment if they recognize themselves and if they have difficult temperaments. We also find sex differences in embarrassment although there are no differences between boys and girls in self-recognition (Lewis, 1992; Lewis, Alessandri, & Sullivan, 1992). Although the term embarrassment has many meanings, here we refer to embarrassment that is due to exposure rather than embarrassment that is due to failure vis a vis a standard. The first type of embarrassment emerges with self-recognition, the second type only after standards and values are established, somewhere around 3 years (Lewis, 1995c). The embarrassment that emerges between 15- and 24-months is related to self-exposure. It is the emotional response when people become the object of others' attention. Most of us think of embarrassment as a failure vis a vis a standard, but there is reason to believe that the early emergence of embarrassment has nothing to do with standards, but with exposure. Even as adults we can be embarrassed without violating a standard. The first example of this type of embarrassment is flattery. Someone is flattered, for example, when they are told how lovely their clothes look, how well they have done their hair, or how pretty the jewelry they are wearing is — a whole series of positive statements — however, that person will become embarrassed. They become embarrassed because they are the object of another's attention.

Differences in showing embarrassment are large and are likely to be related to how the young child responds to being the object of others' attention, which DiBiase and Lewis (1997) have shown are related to difficult temperaments. Some children show pathological embarrassment; however it is often called shyness. Interest in emotional competence involves determining what factors, including socialization factors, are responsible for these pathological responses. Whatever will account for the range of these individual differences, it seems clear that the emergence of embarrassment in the second half of the second year is related to the emergence of an inner life as measured by self-recognition.

Besides embarrassment, empathy also has been related to the emergence of self-recognition. Bischof-Kohler (1991) defines empathy as a child's appropriate facial expression and action in regard to another's distress. She found that the emergence of self-recognition was related to the emergence of empathic behaviour. Her argument is that empathic responses require the child to utilize self-knowledge to "know" about the other's inner state. You cannot place yourself in another's place unless you have an idea of yourself. Thus, the emergence of both empathy and embarrassment are related to self-consciousness.

Finally, the model (see Figure 1) directs us toward the emergence of a new set of emotions at about three years of age. These I have called the self-conscious evaluative emotions. The cognitions necessary for these emotions to emerge are relatively elaborate and require further maturation and socialization. They emerge around three years of age when children are able to evaluate their behaviour in terms of a standard; that is whether their action is successful or not, to feel responsibility for their action, and to be able to focus upon oneself as the source of the failure or focus on the task itself (Stipek, Recchia, & McClintic, 1992). The distinction between focusing on the task or on oneself has been called global vs. specific self-attention (Weiner, 1986). This difference is captured in the adult by such phrases as "I'm no good as a scientist because I had the paper rejected from the journal," as opposed to "Yes, I did not do the proper analysis." The first statement focuses on the self (global), while the second focuses on how to correct the failure (specific). Depending on the focus, different emotions are experienced. In shame, the focus is on a self that is no good. In guilt, the focus is on the action of the self (Lewis, 1992; Tangney, 1995).

We have been trying to measure shame and guilt. Shame is characterized as a collapse of the body. Guilt is associated with the idea of repair and there is no bodily collapse. Barrett (1995), for example, gave children a toy which fell apart as they played with it. Children showed different responses. One child walked away from the toy and did not do anything. It is as if it was not her fault that the toy broke. She took no responsibility. Another child sat and busily tried to repair the doll. A third child hung her head and showed body collapse when the toy fell apart. In the first case, the child does not take any responsibility, in the second she feels guilt, while in the third it is shame that is felt.

In our experimental work with children (Lewis et al., 1992; Alessandri & Lewis, 1993) looking at pride and shame, we have constructed a different task. We gave children matching tasks which required that they match a colored dot to an animal picture. There are simple and difficult versions of this task. In our paradigm, the children have a limited time to finish the task and when the time is up, a bell rings. We are able to manipulate the time so as to produce success or failure as we choose. Each child receives four conditions; an easy task (few dots) which they succeed in doing before the bell rings, an easy task which they fail, a difficult task they succeed in, and a difficult task which they fail.

We have been studying children from less than three years of age through six years. Lewis et al. (1992) studied children at 33 months old. The results of this first study indicated that children almost three years old have the capacity to evaluate their behaviour against some internal standard. For example, when they failed an easy task, they were more likely to show shame than when they failed a difficult task. Likewise, when they succeeded in a hard task they showed more pride than when they succeeded in an easy task. Individual differences were visible and are consistent with the earlier findings with regard to embarrassment. While there were no sex differences in pride, there were large differences in shame. Girls show more shame than boys when they fail tasks.

In more recent work, looking at children's responses from four to six years, we find little age differences in emotional responses to these tasks and to success and failure. Six-year-olds show as much shame and pride as three-year-olds. The lack of age differences reflects the fact that the children of all ages were given the same task. Obviously, if they have different tasks, there might well be age differences since as children get older they are more likely to have more complex standards and more information about what constitutes success or failure. Nevertheless, sex differences remain consistent; girls show more shame than boys while there are no sex differences in pride.

Socialization of Self-Conscious Emotions

Besides sex differences, individual differences in response to failure and success appear in all of the research. In order to investigate where these differences might come from socialization factors were examined; in particular, how parents respond to their children's performances. Recall that these self-conscious emotions require that the child have a standard, feel responsible for their success or failure, and use either self- (performance) or task-focus. Each of these cognitive-attributional attributions involve socialization.

Parents Set Standards. For example, a patient of mine who had extremely high standards related to me the following story. One day he received a grade of 97 on a chemistry exam. When he came home, showing the paper to his father, he reported to me that his father said "What happened to the 3 points?" This example suggests that what constitutes success and failure is likely set by the socializing experience. To begin with, this is likely to be a parent; however, older siblings and friends are also likely to play a role. Some people experience success when they get a C grade and some people experience failure when they get a C grade.

Responsibility for Success and Failure Is Also Socialized. Another patient informed me of a memory he had about responsibility. It involved a time when he spilled his glass of milk. He remembered saying to his mother, "Uh, it's an accident." His mother responded by saying, "Well, if you had done it on purpose, we'd put you in jail." For this mother, accidents were not allowed, the child was always responsible. This, of course, meant that he often felt guilt and shame because he saw himself as responsible even when he was not.

Self or Performance- versus Task-Focus Also Is Socialized. There are data on teachers' behaviour toward students to indicate that this focus can be taught (Deaux, 1976; Dweck & Leggett, 1988; Minuchen & Shapiro, 1983; Nicholls, 1984). For example, when a boy student fails a task, the teacher says, "Let me show you how to succeed. You did it this way, you should have done it that way." In other words, the teacher focuses the boy on the task or on a specific attribution. When a girl fails, the teacher says, "A bad job for a good student." This statement, on the part of the socializing agent, focuses the child on themselves and on their performance.

Finally, Socializing Agents Vary in the Use of Positive and Negative Language and Affect as a Consequence of Success or Failure. For example, in our studies of predominantly middle-class families, approximately 30% of mothers signal their disapproval of their child's behaviour by making a disgust face at least twice during a 15-minute play situation. The consequences of this behaviour are not yet known, but data from Gottman (1979) shows that marital conflict often occurs after one of the adults makes a disgust and contempt face to the other.

There are other socialization experiences which can be interpreted in light of these attributions. Punishment and withdrawal of love are likely to be related to the development of self- versus task-focus while reasoning, which is thought to be an effective socializing practice, is likely to lead to task-focus. Punishment may be related to self-focus rather than to task-focus. Consider the strategy of telling the child they are no good as opposed to explaining why the child failed the standard. If the child is told they are no good, that they are "stupid," there is nothing they can do to change that attribution. If, however, the reason why the child's failure occurred is explained, the child's attention will be focused on the task, not on itself. Love withdrawal as a socialization tool seems to be an extraordinarily powerful technique. From our point of view, love withdrawal focuses the child on self around failure attributions.

In order to explore the problem of self- versus task-focus and its relation to shame and pride, a series of studies were performed in which we were able to identify 4- to 5-year-old children who were self- or performance-oriented as well as children who were task-oriented. These children were given the standard paradigm of easy–difficult, success and failure tasks. As usual, we measured their emotion expressions to success and failure.

Figure 2 presents these data. We predicted that children who were self-focused were going to be more likely to show more embarrassment and shame when they failed and more pride when they succeeded than children who were task-focused. The data for two

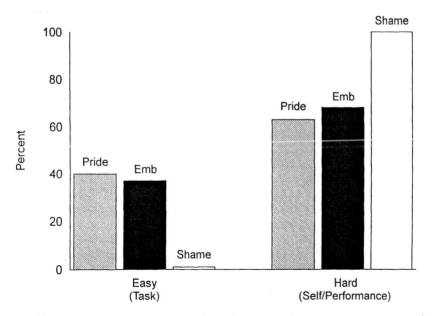

Figure 2. Self-conscious emotional expression (pride, embarassment, shame) as a function of task- versus self/performance-focus.

studies of over 100 subjects show the same findings: The expressions of shame, embarrassment, and pride are much more likely to occur if the child was self- rather than task-focused.

Other data on socialization supports these findings. For example, Alessandri and Lewis (1993) examined the relation between parental statements and children's expression of shame. We found that when mothers use positive statements toward their children's failure, their children show less shame. When they use negative statements, the children show more shame.

Socialization, Maltreatment, and Shame: A Case Study in Risk

Perhaps another way of approaching socialization issues and the development of maladaptive behaviour (incompetence) is to examine what happens to emotional behaviour in children who are maltreated. Alessandri and Lewis (1996) studied a group of children raised in poverty, some of whom were maltreated and some who were not. We tested these children using our standard success–failure paradigm. Because there were important sex differences, boys and girls were considered separately. When girl children failed, they were more likely to show shame if they were maltreated than if they were not. In the case of pride, maltreated girls showed less pride than did the non-maltreated girls. What this suggests is that when girls are abused, they show more shame when they fail and less pride when they succeed than non-abused girls. Perhaps their emotional competence has been negatively affected by their socializing experience.

The maltreated boys showed a different pattern. While boys, in general, show less shame than girls, they usually do not show less pride. Maltreated boys showed less shame when they failed and less pride when they succeeded than did non-maltreated boys and less than did both maltreated and non-maltreated girls. The pattern they showed is a generalized reduction of emotional expression. The effect of maltreatment in boys appears to reduce their emotional expressions, to make them "emotionally numb." What this case study shows is that maltreatment has significant effects on the emotional competence of children. This lack of competence expresses itself very differently for girls and boys. For girls, there is more depression, that is, high shame and low pride, while for boys there is less emotion; they cope with their maltreatment by not feeling. This lack of emotional competence for boys may contain an aspect of sociopathic behaviour.

This way of coping with trauma is consistent with other data on boys' and mens' emotional competence. "Dissing" is a phenomena at the heart of emotional competence. To be dissed means that someone is disrespectful to you. Many years ago, a teacher in a prison came to me with a problem in regard to the boys' aggressive behaviour. In this prison for boys, the inmates would easily punch each other in the face and draw blood, so there was a lot of physical violence. One day, one of the boys "farted" and the boy next to him turned and punched him in the face. Why should one boy punch another for "farting?" We now have a term for this phenomena which grew out of the poverty literature; it is "dissing." The boy who did not "fart" made attributions about the other boy's behaviour. The attribution of dissing led to aggression. As I have tried to indicate (Lewis, 1992, 1995b), the display of aggression is often proceeded by shame (see also Scheff, 1995). In particular, there is an inability to distinguish responsibility and to be so used to failure that many behaviors are misinterpreted as personal failure. What may be happening is that males who are abused or otherwise traumatized are high in shame feeling. To avoid shame, they reduce feeling in general and, when pushed, express anger as a defense to avoid being shamed.

EMOTIONAL BEHAVIOUR AS COMPETENCE

Thus far, I have concentrated on emotional development. As was suspected, there are maturational factors which influence emotional development. Risk to normal maturation, either through inquiry or retardation, results in a loss of emotional competence. Individual differences in the emergence of emotional skills are present as well. These differences have at least two origins and include temperament or dispositional factors as well as socialization factors. Temperament factors involving the lack of arousal or the inability to dampen a response once it occurs, affect emotional expression and regulation. Difficult temperament children are at-risk for emotional incompetence. Socialization factors also play an important role. In terms of the first set of emotions, those called primary or basic emotions, the socialization practices of the caregivers are likely to play an important part in individual differences. Perhaps of even more importance is the role of socialization on the self-conscious emotions. By their very nature, these emotions, requiring elaborate cognitions which include standards and responsibility and evaluation, are most directly influenced by the socializing agents. Thus, any theory of emotional development needs to consider all factors.

Why have we chosen emotional behaviour as an important competence? First, emotions are the content of social exchanges (Lewis & Michalson, 1983) and, thus, play an important role in children's social adaptation. Second, emotional behaviour has been considered as the motives or drives which lead us to learn, to attempt to overcome frustration, and to remember (Tomkins, 1962). For us, emotions, especially those having to do with the self, provide the underfooting for most of the child's developing competence. The failure in emotional development leads to serious implications. For example, school competence is very much influenced by pride and shame. A child who fails and feels shame is a child who will avoid and withdraw from the situation rather than work harder.

The child who cannot cope with shame and instead responds with aggression also shows a lack of competence. The development of emotional competence is a major task since it affects not only emotional life but also our competence in other domains. We need to focus our attention on these tasks, developing more exact models of its development, understanding the role of maturation, temperament, and socialization, and learning how to intervene in order to help children whose development has been deviant. If we can come to understand how emotions develop and how individuals differ, we will have come a long way in understanding the development of competence in children.

REFERENCES

Alessandri, S. M., & Lewis, M. (1993). Parental evaluation and its relation to shame and pride in young children. Sex Roles, 29, 5/6, 335–343.

Alessandri, S. M., & Lewis, M. (1996). Differences in pride and shame in maltreated and non-maltreated preschoolers. *Child Development, 67*, 1857–1869.

Barrett, K. C. (1995). A functionalist approach to shame and guilt. In J. P. Tangney & K. W. Fischer (Eds.), *Self-conscious emotions: The psychology of shame, guilt, embarrassment, and pride* (pp. 25–63). New York: The Guilford Press.

Bischof-Kohler, D. (1991). The development of empathy in infants. In M. E. Lamb & H. Keller (Eds.), *Infant development: Perspectives from German-speaking countries* (pp. 245–273). Hillsdale, NJ: Lawrence Erlbaum.

Bowlby, J. (1969). *Attachment and loss: Vol. 1. Attachment.* New York: Basic Books.

Darwin (1969/1872). The expression of the emotions in man and animals. Chicago: University of Illinois Press. (Original work published 1872).

Deaux, K. (1976). Sex: A perspective on the attribution process. In J. Harvey, W. Iches, & R. Kidd (Eds.), *New directions in attributional research* (Vol. 1). Hillsdale, NJ: Erlbaum.

DiBiase, R., & Lewis, M. (1997). The relation between temperament and embarrassment. *Cognition and Emotion, 11*, 259–271.

Dweck, C. S., & Leggett, E. L. (1988). A social-cognitive approach to motivation and personality. *Psychological Review, 95*, 256–273.

Gottman, J. M. (1979). *Marital interaction: Experimental investigations.* New York: Academic Press.

Izard, C. (1977). Human emotions. New York: Plenum Press.

Lewis, M. (1992). *Shame, The exposed self.* New York: The Free Press.

Lewis, M. (1993a). The emergence of human emotions. In M. Lewis & J. Haviland (Eds.), *Handbook of emotions* (pp. 223–235). New York: Guilford Press.

Lewis, M. (1993b). Self-conscious emotions: Embarrassment, pride, shame, and guilt. In M. Lewis & J. Haviland (Eds.), *Handbook of emotions* (pp. 563–573). New York: Guilford Press.

Lewis, M. (1995a). Aspects of self: From systems to ideas. In P. Rochat (Ed.), *The self in early infancy: Theory and research* (pp. 95–115). Advances in Psychology Series. North Holland: Elsevier Science Publishers.

Lewis, M. (1995b). Embarrassment: The emotion of self exposure and embarrassment. In J.P. Tangney & K.W. Fischer (Eds.), *Self-conscious emotions: The psychology of shame, guilt, embarrassment, and pride* (pp. 198–218). New York: Guilford Press.

Lewis, M. (1995c). Self-conscious emotions. *American Scientist, 83*, 68–78.

Lewis, M. (1995d). *Shame: The exposed self* (paperback edition). New York: Free Press.

Lewis, M. (1997). *Altering fate: Why the past does not predict the future.* New York: Guilford Press.

Lewis, M., Alessandri, S., & Sullivan, M. W. (1992). Differences in shame and pride as a function of children's gender and task difficulty. *Child Development, 63*, 630–638.

Lewis, M., & Brooks-Gunn, J. (1979). *Social cognition and the acquisition of self.* New York: Plenum.

Lewis, M., & Michalson, L. (1983). *Children's and moods: Developmental theory and measurement.* New York: Plenum.

Lewis, M., & Ramsay, D. S. (1997). Stress reactivity and self-recognition. *Child Development, 68*, 621–629.

Lewis, M., & Ramsay, D. S. (in press). Onset of self-recognition and the emergence of self-awareness. *Developmental Psychology.*

Lewis, M., Sullivan, M. W., Stanger, C., & Weiss, M. (1989). Self-development and self-conscious emotions. *Child Development, 60*, 146–156.

Minuchen, P. P., & Shapiro, E. K. (1983). The school as a context for social development. In P. Mussen & E. M. Hetherington (Eds.), *Handbook of child psychology* (4th ed., Vol. 4, pp. 197–274). New York: Wiley.

Nicholls, H. G. (1984). Achievement motivation: Conception of ability, subjective experience, task choice, and performance. *Psychological Review, 91*, 328–348.

Prigatano, G. P. (1991). Disturbances of self-awareness of deficit after traumatic brain injury. In G. P. Prigatano & D. L. Schacter (Eds.), *Awareness of deficit after brain injury* (pp. 111–126). New York: Oxford University Press.

Scheff, T. J. (1995). Conflict in family systems: The role of shame. In J. P. Tangney & K. W. Fischer (Eds.), *Self-conscious emotions: The psychology of shame, guilt, embarrassment, and pride* (pp. 393–412). New York: The Guilford Press.

Stipek, D., Recchia, S., & McClintic, S. (1992). Self-evaluation in young children. *Monographs of the Society for Research in Child Development, 57*(1), (Serial No. 226).

Tangney, J. P. (1995). Shame and guilt in interpersonal relationships. In J. P. Tangney & K. W. Fischer (Eds.), *Self-conscious emotions: The psychology of shame, guilt, embarrassment, and pride* (pp. 114–142). New York: The Guilford Press.

Tomkins, S. S. (1963). *Affect, imagery, and consciousness: Vol. 2. The negative affects.* New York: Springer.

Weiner, B. (1986). *An attributional theory of motivation and emotion.* New York: Springer-Verlag.

3

SOCIOEMOTIONAL AND COGNITIVE COMPETENCE IN INFANCY[*]

Paradigms, Assessment Strategies, and Implications for Intervention

Dale M. Stack and Diane Poulin-Dubois

Centre for Research in Human Development and Department of Psychology
Concordia University

COMPETENCE IN INFANCY

It was not so long ago that infants were considered passive recipients of their external world, receiving and processing information as it impinged on them, without being considered to have any control over the world themselves. Similarly, William James (1890) referred to babies as living in "buzzing, blooming confusion." However, Piaget (1952) revolutionized thinking about early development, describing some of the strikingly sophisticated mental capacities of the developing human infant and revealing what was in his view the constructive and active, human mind. His and current views of infants have discarded "passivity" and reactivity for activity and initiative, active participation, and competence. Developmentalists hold and some even embrace, bidirectional and interactive models, or alternatively, dynamic and complex systems models, and recognize how infants exert control on their physical and social environments even early on, and how they learn to relate to their world. The infant actively participates in a changing social context. Development is dynamic and changing, and the changing nature of both the child's environment and biological and constitutional factors, as well as, its relationships are considered important contributors to be studied over time (Sameroff, 1975). In this "transactional" (Sameroff & Chandler, 1975) and "dynamic" (Fogel, 1993) view, it is critical to study how transactions between infant and environment facilitate or hinder adaptive integrations and adaptations as both infant and surroundings develop and evolve, in order to understand developmental processes.

[*] This chapter was written with the support of Conseil Quebecois de la Recherche Sociale and Natural Sciences and Engineering Research Council of Canada. The authors thank Nadine Girouard, Mandy Steinman and Barbara Welburn for their help with preparation of the final version of the chapter, and Dolores Pushkar for editorial suggestions. Address correspondence to the authors at Department of Psychology/Centre for Research in Human Development, Concordia University, 7141 Sherbrooke St. West (PY-170), Montreal, Quebec, Canada H4B 1R6.

Improving Competence across the Lifespan, edited by Pushkar *et al.*
Plenum Press, New York, 1998.

While infants' competencies are becoming widely disseminated, their vulnerabilities are nonetheless also salient. While at first glance this statement may appear to be a paradox, both competencies, as well as protective factors or inherent self-righting principles in development, are evidenced in the research literature (e.g., Rutter, 1987; Coie et al, 1993). Human infants are one of the few species to spend a 9-month gestational period only to spend many more years of their infant and child life supported and nurtured by adults, without whose care they would not survive. While not necessarily predisposed, infants are susceptible to numerous genetic, biological and environmental factors (both negative and positive), prevalent abuse and neglect, and can experience a variety of adaptive difficulties. However, whether experiences suffered during infancy have long term negative impacts on later development is controversial.

During the 1960s and '70s, the emergent interest in infancy research fostered a belief that infants' experiences must have a powerful effect on human development (Lamb & Bornstein, 1987). At that time, the critical formative importance of early experiences was also inherent in several leading theories (e.g., learning, psychoanalysis). In his review of the status of research in infancy for the Grant Foundation, Kagan (1982; cited in Lamb & Bornstein, 1987) concluded that experiences in infancy frequently play only a minimal role, if one at all, in shaping later development. He contended that discontinuities rather than continuities in development were common. The truth is that predictability from infancy to later performance is poorest relative to later ages (Kagan, Kearsley, & Zelazo, 1978; McCall, 1979). For example, studies of extreme environmental circumstances in infancy, such as the Romanian experience (e.g., Ames, 1997 final report), have shown the extent to which some of these traumatic events can be overcome (e.g., Rutter, 1981). Such "resiliency" characterizes early development and is related to the fact that there are often several alternative pathways to achieve the same developmental goal. Bornstein and Lamb (1992) point out correctly, however, that while not all infant experiences are critical for later development, and not all experiences are equal in salience or impact, some infant experiences do have lasting effects.

It is nonetheless clear that infants' experiences are important because to some extent the later environment depends on what has gone on before. The history of the individual is significant and psychological functioning can be impacted by acute stressors and chronic adversities (Rutter, 1987). However, it is also worthy of note that these environmental and biological insults impinge on an active organism, and there can be marked individual differences in response owing to, for example, differences in susceptibility, different contexts, levels of resilience, and in coping and adaptation (Rutter, 1987). That is, differences in the person–environment interactions. This individual difference issue raises the question of intervention and how to know when, and in what cases, to intervene.

Indeed, while the ability to overcome adverse outcomes in some cases is a clear illustration of a hopeful picture, it also presents a major challenge at the same time. As Emde asked, "how are we to prevent later maladaptive outcomes through early intervention efforts, and how are we to be assured of any continuity from our efforts?" (1987, p. 1301). The question seems clear, although the solution is neither simple nor easy. Competence is dependent on context and both endogenous and exogenous factors play roles. However, the effort is not hopeless — one key player in discriminating competence from incompetence that warrants increased attention is assessment. Valid assessment instruments are essential in order to identify problems early, as a means of monitoring problems as they develop, and of observing changes as they occur. However, at this point in time our knowledge is limited about how to ascertain and discriminate competence versus incompetence in the early years, and assessment strategies and intervention efforts to this end are thus both complicated to design and sparse.

EARLY ASSESSMENT AND INTERVENTION

A standard definition of assessment is "the measurement of the relative position of an individual in a larger defined population with respect to some psychological construct" (Cicchetti & Wagner, 1990, p. 248). An adequate assessment tool is required to have reasonable external validity, as well as, tap a construct that is stable over time. As we will see later, these two basic requirements have not been fulfilled by traditional assessment batteries of early intellectual functioning. Consequently, prediction has been difficult. The establishment of the position of an individual within a group is another reason why assessment is important as it provides the basis from which one can monitor the effects of intervention (Cicchetti & Wagner, 1990). Again, traditional assessment tools have proven unsatisfactory in this regard as two children with the same IQ scores do not necessarily have the same competence in the same specific cognitive domains. This lack of specificity impedes the practitioner from devising appropriate intervention programs for a given child.

The historical roots of the early childhood intervention movement extend back to the early twentieth century, and perhaps further. A number of domains, including early childhood education, maternal and child health, special education, and child development research, are considered sources from which the roots of early intervention derived (Shonkoff & Meisels, 1990). Of the themes developed over the early years, several have persisted and are reflected in current thinking and intervention efforts (Shonkoff & Meisels, 1990). These persistent themes include a commitment to the special needs of children recognizing their vulnerability (whether that vulnerability be due to biological or disabling conditions versus conditions of poverty in their environment for growth), a belief that society is responsible for the care and welfare of children, and a view that prevention is better than treatment, and early intervention a better approach than later remediation (Shonkoff & Meisels, 1990).

The provision of enrichment and stimulation to infants who were at biological or environmental risk for developmental problems (Tjossem, 1976, cited in Black, 1991), and closing the gap between disadvantaged and more advantaged children (Benasich, Brooks-Gunn, & Clewell, 1992) were the noble goals of early intervention programs. These were accomplished in part by increasing competence in social and educational domains. There are numerous examples of early intervention programs (e.g., Head Start, Zero to Three, Mother–Child Home Project, Prenatal/Early Infancy Project, Project CARE, etc.), and these vary along a number of dimensions including the groups they target (e.g., disadvantaged, preterm, developmentally disabled). According to Benasich et al. (1992), intervention can occur at the level of individual, dyad, or group, and a variety of strategies and protocols have been used, depending on the objectives of the program. Intervention programs vary on the basis of service delivery context (home, school, clinic), principal target (mother, child, or both), commencement point (prenatal, infant, toddler), intensiveness (amount of programming), length/duration of the program, content of the curriculum and program, number of staff to child ratio, and training (Benasich et al., 1992). While all programs appear to have had the general objective of enhancing child competence, the specific foci have often varied, from emphasizing the mother–infant relationship or their interactions, to health, language, home environment; the most common area targeted has been child cognitive outcome (Benasich et al., 1992).

Early intervention programs were traditionally premised on the belief that those children faring poorly early in life were anticipated to remain as such (Sameroff & Fiese, 1990). These stable models of development are exemplified in the Head Start program which was designed to provide stimulation and learning and enhance self-concepts in preschool children with the goal of having the gains and successes achieved continue into later life (Sameroff & Fiese, 1990). However, only moderate gains in intellectual competence were maintained, although school failure rates and the need for special education were reported to have been reduced (Sameroff & Fiese, 1990). Similarly, the majority of those children identified as at risk due to biological conditions did not show the negative outcomes that were generally expected to be the case. These two findings, based on numerous studies, suggested that early characteristics of the child might interact with environmental context. Sameroff and Fiese (1990) drew two conclusions on the basis of these findings: (1) level of competency as a child does not reflect a linear relationship to competence in later life; and, (2) the influence of the child's social and family life must be included in any predictive equation, as these factors may promote or hinder a positive developmental course. The findings also suggest, however, that more focused, sensitive assessment tools are necessary. A more comprehensive and better assessment facilitates determination of both type and intensiveness of intervention.

Consequently, in attempts to explain and prevent problems of human adaptation, it is essential that emerging themes from the study of human development be integrated (Coie et al., 1993). This would include concerted efforts to define and identify problems in early infancy, as well as, intervention and prevention strategies to meet the needs. Toward this end, we believe that powerful paradigms and theories developed from research in infancy have important implications and applications to early intervention, and have already to date, proven influential. The present chapter seeks to elucidate the importance of these directions, particularly as they relate to infants' socioemotional and cognitive development, and to highlight the challenges they present and the future avenues that become essential to pursue.

ASSESSMENT OF SOCIOEMOTIONAL COMPETENCE

Much of children's socioemotional development takes place within the context of the parent–child relationship and the environmental milieu within which they live. Moreover, the building blocks for social competence and resilience are laid in infancy. However, much remains unknown about developmental processes in social and emotional development in infancy, and there are few comprehensive theories to guide the research, particularly in emotional development (e.g., Lewis, 1996). During the first year of life, there is ample evidence that many of infants' perceptual and cognitive abilities are used to understand their social environment (Lewis, 1987) and data indicate that infants are capable of differentiating between social and nonsocial dimensions (Lewis, 1987; Poulin-Dubois, in press). Similarly, the importance of the early caregiver–infant relationship at this developmental stage has been well documented. During this stage, infants' abilities to communicate and regulate emotion grow, and they learn to distinguish and respond appropriately to the emotional signals of others (Segal, Oster, Cohen, Caspi, Myers, & Brown, 1995). Communicating information about their needs, goals, and affective responses to social and environmental stimuli is crucial for infant development. The capacity to modulate these signals and use them effectively unfolds within the context of repeated interactions with parents, i.e., within the context of the developing relationship (Segal et al., 1995; Fogel, 1997). This link to interactions is being reflected in more clinical and remediation approaches taking the family environment into account, advocating that the relationship is the appropriate context (e.g., Emde, 1987; Fogel, 1993). Consequently, it is essential to underscore the importance of assessing parent–child relationships when conducting assessments of the emotional development of infants and toddlers (Cicchetti & Wagner, 1990).

Directly assessing the child's socioemotional functioning is as important as assessing other developmental areas (Denham, Lydick, Mitchell-Copeland, & Sawyer, 1996). Socioemotional variables may also be considered indices of adjustment, and according to Lazar and Darlington (1982, cited in Denham et al., 1996) these are often associated with cognitive changes and may be more sensitive to the effects of intervention. The abundance of evidence notwithstanding, such assessment measures remain scarce. Those measures that do exist are less refined and valid relative to those that exist for cognitive development. However, there are a number of existing measures and several research procedures that shed light on socioemotional development and hold promise for assessment depending on how they evolve and the extent to which they are validated as assessment protocols for use in clinical settings. We begin by covering laboratory based procedures, followed by a sample of observational home-based measures, and end with several more norm referenced assessments, as well as, examples of promising new measures.

Socioemotional development and the quality of the mother–child relationship can be examined by studying face-to-face interactions between caregiver and infant over the first 6 to 12 months of life. Face-to-face interactions have been one of the primary means used to study the infant's social communication (Kaye, 1982), emotional expressions and responses to stressful episodes (Field, Vega-Lahr, Scafidi, & Goldstein, 1986), and development of social expectations (Cohn & Tronick, 1983). These interactions can shed light on important features of the developing relationship. If interactions are disrupted or disturbed, the underlying relationship may be affected (Field, 1987). Researchers have begun using these procedures to study mothers and infants who are at risk for affective and interactive disturbances. It is believed that factors that interfere, on the one hand, with the infant's capacity to modulate its signals, communicate or regulate its emotions, or, on the other hand, with the caregiver's furnishing of sensitively timed and appropriate responses, can impair interaction quality or modulate the development of the relationship, and impede progress in communication skills (Ainsworth, Bell, & Stayton, 1974; Isabella, 1993; Segal et al., 1995). Interactions of high-risk mother–infant dyads are often lacking in affective displays and contingent responsivity, and the behavior patterns are not synchronous (Field, 1987). According to Field, the potential consequence when mothers are insensitive to timing or unresponsive to infants' cues is that infants may become distressed or withdraw from the interaction. Alternatively, there is some evidence to suggest that when mothers are affectively unresponsive (e.g., depressed), little agitation in the infant is apparent, but the infant tends to mirror mother's lack of affective activity (Field, 1987). In either case, the outcome is aberrant.

During these face-to-face interactions, infant and adult (primarily the mother) are seated at eye-level to each other for a series of brief interaction periods. Mothers interact spontaneously, using facial, vocal and tactile expressions, while infants respond to and even initiate interactions. Perturbations and manipulations of these face-to-face interactions have also been successfully used. Free play interactions on the floor of the home or clinic setting provide another means of observing the dyad, and allow for objects to be a part of the play situation. Several different types of observational coding schemes can then be applied, e.g., Field's Interaction Rating Scales (1980) which is a judgment-based measure, or multiple measures from mother and infant coded.

The still-face (SF) procedure (Tronick, Als, Adamson, Wise, & Brazelton, 1978) is a modified face-to-face procedure, where the mother–infant interaction is divided into three brief face-to-face periods. In periods 1 and 3, the mother interacts normally, using facial expression, voice, and touch (Normal), while in period 2 she assumes a neutral, non-responsive still face and provides neither vocal nor tactile stimulation (SF). Infants typically react to the SF with reduced gaze and smiling, and increased neutral/negative affect, with accompanying adaptive strategies to engage and regulate themselves. By manipulating maternal behavior in the SF, it has consistently been shown that infants develop expectations about their mothers' behavior during social interactions, that the procedure taps meaningful individual differences in the quality of the developing mother–infant relationship, and that it reflects the history of the interaction, not merely infant characteristics (e.g., Cohn, Campbell, & Ross, 1991). Reliable differences between various risk groups (e.g., depressed and cocaine abusing mothers) have been found. Furthermore, by comparing a standard SF with one where mothers can touch during the SF period, it has been demonstrated that touch can modulate infants' responses to the SF, by eliciting high levels of gaze and smiling, with little distress (Stack & Muir, 1990). Given that the amount and quality of physical contact (touching, proximity) are also important to the mother–infant relationship (Montagu, 1986; Stack & Muir, 1992; Ainsworth, Blehar, Waters, & Wall, 1978), and that the importance of touch in intervention programs with preterm infants has been demonstrated (e.g., Scafidi et al., 1986; Rausch, 1984) pursuing these avenues appears promising.

Taken together, these studies show promise for face-to-face and SF procedures in revealing socioemotional indices and markers, and highlight the importance of mothers' and infants' sensitivity to each others' behavior. Sensitivity and responsiveness have direct links to attachment, development of the relationship quality, and future interactions.

Probably the most well known laboratory measure that assesses emotional signalling and caregiver–infant availability, and evaluates the caregiver–infant quality of attachment is the "strange situation" (Ainsworth & Wittig, 1969). A series of brief observations of play and separations and reunions with mother were designed to assess security of attachment and secure-base behavior in infants and toddlers. It is not, however, a diagnostic instrument. The strange situation has generated a number of attachment profiles and has been an important catalyst for researchers to develop additional age-appropriate assessments in older children. However, the procedure relies on the assumption that it actually activates the attachment system (Seifer & Schiller, 1995). As Seifer and Schiller point out, along with normative descriptions of secure-base behaviors come hypotheses regarding the development of individual differences. Taken in this context, they argue that constructs of parenting sensitivity, consistency, parents' cognitive constructions of their own relationship history, child temperament, and infants' signalling behaviors make important contributions. Two other observational assessments include Lewis and Michalson's (1983) scales of socioemotional development, and Stern's (1985) systematized observations of "affect attunement." While these latter measures also show promise, they have been less widely used (Emde, 1993).

Given the central role that the mother–infant relationship plays in a child's social, emotional, and cognitive development (Bretherton, 1987; Field, 1987; Ricks, 1985), observational, rather than laboratory, measures which assess the caregiver–child relationship are also important. One of the most extensively used observational methods to assess caregiving and problems in caregiving has been the Ainsworth Sensitivity Scales (Ainsworth et al., 1978). They were designed to be used in the home and observe spontaneous interactions. Biringen, Robinson, and Emde (1988) developed the Emotional Availability Scales (EAS) in an effort to modify the Ainsworth scales. They reconceptualized emotional availability as a relational construct that includes the integration of different components of maternal and child behavior that are present in an early relationship. The EAS are global rating scales designed to assess the quality of the mother–child interaction as it occurs during free play, for example. The scales are theoretically based and tap dimensions of maternal and child behavior. They have been used to assess the quality of the mother–infant emotional communication in both normal and risk samples with children from 1 to 8 years (e.g., Easterbrooks, Lyons-Ruth, Biesecker, & Carper, 1996).

Three other measures worthy of mention are the Greenspan-Lieberman Observation System for assessment of caregiver–infant interaction during semi-structured play (GLOS; Clark, Paulson, & Conlin, 1993; Greenspan, 1983; Greenspan & Lieberman, 1989), the Parent–Child Early Relational Assessment (PCERA; Clark, 1985; Musick, Clark, & Cohlen, 1981), and the Child–Adult Relationship Experimental Index (CARE-INDEX; Crittenden, 1981). The PCERA was developed in the context of the Mothers' Project and has largely been used with severely disturbed families, while the CARE-INDEX was developed largely for research purposes, is limited to the younger age range of infancy, and has not been examined for its suitability for applied use.

The importance of studying emotions (and as a result, the necessary extension to assessing and intervening at the level of emotions) is captured in part in Lewis's (1992) argument that it is the early years that are critical for emotional and social growth and health. Lewis and his colleagues (Lewis, 1992; Lewis, Alessandri, & Sullivan, 1992; Alessandri & Lewis, 1993) have designed a number of promising laboratory procedures to assess the sub-conscious emotions (e.g., shame, guilt, embarrassment, pride) and the role(s) that parental and child evaluations play. To assess shame and pride, for example, parent–child dyads are engaged in both a free play and a problem solving situation. The problem solving task is structured and consists of easy and difficult problems to solve, where a timer is used and no assistance is provided. Explanations and behavioral responses (both bodily movements and facial expressions) to success and failure as a function of task difficulty are used as indices of emotional behavior. Typically, more shame is expected when failing the easy tasks and more pride when successful at the difficult tasks. The organization of the sub-conscious emotions over task, situation, parent, and gender, as well as documenting the developmental course of these emotions and their meaning for individual differences, holds promise for revealing a new means of assessment.

There are several measures that are more norm referenced and assess, to some extent, socioemotional domains. The Neonatal Behavior Assessment Scale (NBAS; Brazelton, 1984) covers the period from birth to one month, and assesses a broad range of neonatal behaviors of which socioemotional domains are some. It evaluates the infant's ability to respond to both examiner and objects during a standardized 20 to 30 minute interaction. The Behavior Rating Scales (BRS) of the Bayley Scales of Infant Development (BSID-II; Bayley, 1993) are norm-referenced scales covering a broader age range relative to the NBAS, and assess a number of relevant domains such as attention/arousal and orientation/engagement. However, they were not necessarily developed to assess socioemotional domains and to date there is little data on their validity and predictive value. A more specific measure to document affect development is Sroufe and Wunsch's (1972) standard protocol of 30 items used to assess the development of smiling and laughter in normal infants from 4 to 12 months of age. It is conducted in the home during monthly visits.

Finally, we turn to two additional measures which offer some promise and have specific purposes. First, the IFEEL pictures is an instrument for interpreting emotions (Emde, Osofsky, & Butterfield, 1993); more specifically, parental interpretation of infant emotions. Parents are shown a standard set of 30 infant photographs and asked to assign "the strongest and clearest feeling that is expressed by the baby." There are typically a number of different emotions that can be used to characterize any given picture, however, an overall profile of emotionality of the respondent is presumably projected. Both a categorical and a dimensional method can be used to score the responses from IFEEL pictures (Butterfield & Ridgeway, 1993). A number of studies with different groups have been conducted including adolescent mothers (Osofsky & Culp, 1993), mothers at risk for child maltreatment (Butterfield, 1993), depressed mothers (Zahn-Waxler & Wagner, 1993), and mothers of premature babies (Szajnberg & Skrinjaric, 1993). While promising as a measure of emotional availability, this picture-based instrument warrants more research and validation before its potential and clinical utility can be assessed with certainty.

Second, Kagan and his colleagues have developed a laboratory procedure to assess temperament, and specifically, inhibition, in infants and toddlers. There is some support for the notion that long term studies of temperament predict to later personality (Caspi & Silva, 1995), that temperament emerges early in development and is defined by "inherited coherences of physiological and psychological processes" (Kagan, 1994, p. 35), and that membership in a temperamental category implies a bias in favor of certain affects and actions (Kagan, 1994). Kagan's procedure was developed around the notion of whether infant behavior could predict inhibited and uninhibited profiles in the second year. Combinations of multiple behavioral and physiological response measures are used at several points across the first few years of life under a variety of conditions which provoke reactivity (e.g., moving mobiles, cotton swab to infant's nose, encountering an unfamiliar object or adult, application of electrodes and BP cuff, atypical requests by strangers). This procedure is both original and enlightening, however, its implications for socioemotional assessment are still to come.

There are also new directions being revealed in nonverbal communication which have implications for socioemotional development. For example, Mundy and Willoughby (1996) contend that nonverbal referential and social communication skills may be important markers related to later language problems and socioemotional difficulties. One route of pursuit emphasizes the development of joint attention skills and their role in the development of "affective intersubjectivity" (Mundy, Kasari, & Sigman, 1992). Thus, it may be that observations of nonverbal communication can provide important information.

Summary

The limits and dearth of available tools for socioemotional assessment purposes have made it difficult to identify infants in need of services and have also limited progress in intervention (Denham et al., 1996). The period of infancy becomes particularly salient in this regard because examining and being knowledgeable about precursors of emotional difficulties in infants is important for prevention and success of intervention. Moreover, it is not sufficient to merely evaluate the child; there may be relationship difficulties and/or caregiving issues at the root of the problem, contributing to the problem, or exacerbating the problem. Defining social and emotional competence and what this competence means at each age of the preschool periods is important if we are to design better assessment tools and intervention strategies and evaluate them (Denham et al., 1996). It is reasonable that there should be assessment procedures appropriate to each age period given that each age period has salient issues and key developmental tasks pertinent to socioemotional development (Denham et al., 1996; Howes, 1987; Waters & Sroufe, 1983). Similarly, Cicchetti and Wagner (1990) contend that the emergence of stage salient tasks (defined as "those developmental issues that are most critical for the adaptive functioning of the child at a given point in time, and that require coordination of affect, cognition, and behavior," p. 263; Sroufe, 1979) are worthy of examination in order to better understand socioemotional development.

Given that assessment measures are scarce, it is not surprising that interventions at this level are practically non-existent. Intervention efforts and even most examples of infant resiliency concern cognitive functions, and few have dealt with social and affective functions. Although intervention often targets self-esteem and confidence, social skills, and emotional development, the interventions are not always specifically designed for these domains, but rather, these are expected outcomes from more general interventions. Moreover, socioemotional goals in intervention are often implicit rather than explicit (e.g., Denham et al., 1996). A clear need to pursue the development and validation of assessment tools for socioemotional development and social competence is underscored, however, this is not sufficient. Rather, efforts to translate and extend assessment into clear intervention goals and programs are warranted. This being said, these efforts must be cautious, recognizing how little we know about prediction, as well as, respecting the importance of individual differences.

ASSESSMENT OF COGNITIVE COMPETENCE

Unlike the unvalidated instruments or tasks available to assess infants' socioemotional competence, standardized normative tests have been used as outcome measures of intelligence for a long time. However, these traditional psychometric tests to assess infant intelligence, such as the Bayley Scales of Mental and Motor Intelligence (Bayley, 1969) largely tap sensory and motor capacities. They were developed following the influence of Gesell who produced an adaptation of the Binet test for the assessment of infants. While Piagetian theory revolutionized our views of infant development, it did not detract from the emphasis on the sensorimotor competence of the infant (Lecuyer & Streri, 1994). In part, because of this emphasis on sensorimotor skills, standardized infancy measures have been plagued by the failure to reliably predict later childhood IQ (see Kopp & McCall, 1982, for a review). In a review of over 20 studies, Fagan and Singer (1983) reported an average correlation of 0.11 between early and later IQ scores for normal populations and somewhat larger correlations with high-risk or handicapped populations. This lack of stability has been explained by a shift in the nature of intelligence from infancy to childhood (McCall, 1979), or by the genetic determinism of early intelligence (Fishbein, 1976; Vernon, 1980). Moreover, several investigators of infant cognitive development have argued that the low predictive power of the traditional infant tests stems from fundamental flaws in the infant tests themselves (Bornstein & Sigman, 1986; Rose, Feldman, & Wallace, 1988). It has been suggested that memory and perceptual skills typically included in later intelligence tests are strikingly absent in infant tests. A need for measures that assess purely cognitive functioning and performance in infancy is warranted (Lewis, 1971; McCall, 1981).

A further limitation of the prevailing standardized assessment procedures is that the cognitive abilities of infants with delays in expressive behaviors are underestimated. As pointed out by Zelazo (1988), the vast majority of children with developmental problems, such as neuromotor problems, expressive language delays, and behavior problems, are the children who need intellectual assessment most. However, delayed motor development severely limits performance on conventional tests, which rely heavily on motor skills such as imitation and receptive and/or expressive language skills. The ability to comply with the examiner's requests is another requirement of conventional tests. Lack of compliance might convey an impression of mental deficiency when the child might be intellectually competent. Alternate paradigms for assessing perceptual–cognitive capacities are therefore desperately needed; in our view, these paradigms will most likely come from recent procedures designed to learn about the cognitive development of normal infants. In the present section, we examine the actual and potential applications of some of the newest research paradigms in infant cognition.

Our knowledge about infants' cognitive competence has increased dramatically over the last few decades. This increased knowledge stems from both theoretical and methodological contributions. At the theoretical level, the Behaviorist Infant has been followed by the Piagetian Infant, the Nativist Infant, and, more recently the Connectionist Infant (Karmiloff-Smith, 1996). At the methodological level, the explosion of research on infant cognition is partly due to the development or adaptation of paradigms that have opened up new possibilities. These include the preferential looking, the habituation/dishabituation, the familiarization–novelty, and conditioning paradigms (Cicchetti & Wagner, 1990). Theoretical and methodological advances have led to a view of the infant as much more cognitively competent than was the case only 30 years ago. It is now established that infants develop sophisticated knowledge about object action and human behavior by the end of the first year (Spelke, Phillips, & Woodward, 1995; Poulin-Dubois, in press). Basic research into the cognitive capacities of human infants represents one of the most promising avenues for the development of new, nontraditional assessment tools (Cicchetti & Wagner, 1990). The time is ripe to examine the benefits (or lack thereof) that the recent advances in research on infant cognition have had on the development of assessment tools to detect nascent psychopathology. For example, recent decreases in mortality rates of low-birth-weight infants have stimulated interest in the developmental outcomes in this population (Kopp, 1987).

An information-processing orientation to the assessment of cognitive competence emerged in the 1970s (Klahr, 1993). This approach is premised on the basic assumption that human children, like adults and other animals, process (e.g., select, encode and remember) information. Attention is a basic component of cognitive functioning and measures such as decrement and recovery of attention are considered to capture information processing reliably during the first months of life. Cross-modal transfer is another cognitive skill that has been intensively investigated (Rose & Ruff, 1987). It reflects the infant's ability to evaluate and compare information across different sensory modalities. To assess these cognitive skills in normal infants, several procedures, such as the preferential looking, the habituation, and the familiarization–novelty paradigms, have been developed which have changed our way of studying infant perception and cognition. In the preferential looking paradigm, a standard procedure consists of the presentation of two visual or auditory displays that differ on some predefined feature or in their consistency with a previous display. In the habituation paradigm, a display is repeatedly presented, followed by either a new target or the same target after the infant's attention has declined to a predetermined percentage of what it was initially. In the case of the familiarization-novelty paradigm, the infant is presented with a stimulus for a number of familiarization trials, followed by a test trial in which the familiarization stimulus is paired with a novel stimulus. These fundamental paradigms have been applied successfully to answer numerous questions about early perceptual–cognitive development.

The critical question is whether these paradigms have any clinical relevance. Normal infants clearly vary in their attentional abilities by the middle of the second year (Colombo & Fagan, 1990). Meta-analyses of the relation between individual differences in the expression of these abilities and cognitive performance in later childhood have revealed correlations that are quite robust across a large set of variables, including various novelty tasks (Bornstein & Sigman, 1986; McCall & Carriger, 1993). Some have argued that these results have more theoretical than practical importance because the size of the correlation (.45) is still too modest for the diagnosis of individuals and is not higher than that for SES (McCall, 1994). At a theoretical level, rate of information processing has been suggested as the underlying stable process mediating mental development (McCall, 1994).

Despite the improved but still modest predictive power of selective attention measures in infancy, the urgent and increasing need to identify those infants at higher risk of cognitive deficits or delays (e.g., preterm, low-birth-weight, physically handicapped) has stimulated the development of standardized tests of infants' information processing skills (Fagan & Shepherd, 1987; Zelazo, 1988). For instance, measures of visual recognition have been found to discriminate among groups of infants expected to differ in intelligence later in life, including infants with Down Syndrome, and low-birth-weight, preterm, and failure-to-thrive infants (Benasich & Bejar, 1992).

The Fagan Test of Infant Intelligence (Fagan & Shepherd, 1987) is a standardized procedure that was developed from the novelty preference paradigm. It is based on the assumption that infants tend to differentially fixate novel, as compared to previously seen, visual stimuli (Fagan & Detterman, 1992). The infant is presented with a standard series of pictures for a predetermined period of time followed by a pair composed of the familiar picture and a novel picture. A novelty preference score is computed on the basis of the amount of fixation during the test phase devoted to the novel picture divided by the total fixation time to both the novel and familiar pictures. The test, if administered repeatedly during infancy, apparently provides a high prediction to later intellectual deficit or normality (McCall & Carriger, 1993).

Despite their many advantages over the traditional tests, the validity and appropriateness of measures based solely on visual attention have been challenged. The limitation of visual fixation as a single measure, the inferential process required by the habituation and paired comparisons, and the reliance on static visual stimuli are some of the shortcomings which have been identified (Zelazo & Stack, 1997). A paradigm based on rate of information processing was developed by Zelazo (1979, 1988) in order to deal with some of these shortcomings. This Standard-Transformation-Return (STR) procedure represents an attempt to overcome the limitations of the tests based exclusively on visual fixation and static displays by measuring clusters of expressive behaviors elicited by infants (e.g., smiles, vocalizations, pointing) in combination with high levels of attention and heart rate decrement to both sequential visual and auditory events. The STR paradigm involves the repetition of a standard to build an expectation, transformations of the standard to assess reactions to novelty, followed by the reappearance of the standard. The child's capacity to create mental representations for events and to measure the rate at which these representations are formed is tested. The procedure has proven useful in identifying children with intact processing ability despite delays on conventional tests due to limited motor and/or expressive skills (Zelazo & Stack, 1997).

Although intervention programs directly derived from the information processing theoretical framework have not yet been developed, these measures have already proven helpful in screening children who require treatment from those who do not. In one study, information processing procedures were used to identify children with intact processing ability from among a sample of children with delays on conventional tests (Zelazo, 1988). Parent-implemented treatment designed to stimulate expressive language and motor skills and to eliminate noncompliant behaviors were offered to 54 children aged 22 or 32 months with developmental delays. Follow-up administration of the Bayley or Stanford-Binet took place 6 and 18 months after the treatment phase. The prediction that children whose information processing ability appeared age-appropriate would improve over test-ings, whereas children with impaired processing ability would not, was supported. Although the results of this study demonstrate the limitations of the traditional tests and the effectiveness of the parent-implemented treatment procedures, they provide less information about the type of treatment that would be required to improve the deficient processing ability of the truly delayed children. It is hoped that the development of such intervention programs will appear on the agenda of infancy researchers in the near future.

The development of assessment paradigms rooted in an information-processing approach was stimulated in the 1970s by the explosion of research in infant cognition in the preceding decade. Since then, research on early infant cognition has expanded in many ways. Scanning the research of the last 20 years in search of new paradigms that might replace or complement those based on the information-processing approach might prove valuable. At first glance, no revolution of a magnitude similar to the one which took place in the late 1960s can be identified. However, as mentioned before, a plethora of models of cognitive development has characterized the past two decades (e.g., Dynamic Systems Theory, Nativism, Connectionism, Social-Pragmatism, Neo-Piagetian). Some of these models recognize the transactional nature of early cognitive development, including the acquisition of language (Tomasello, Kruger, & Ratner, 1993). Although most of these theoretical contributions have not been concerned with individual differences and the issue of predictability, there are indications that some of the paradigms used to test their predictions (e.g., imitation, joint attention) have the potential to reveal individual differences of interest to the clinician looking for assessment instruments. As well, some of these paradigms consider the caregiver–infant dyad in assessing infants' intellectual skills. We now review some recent empirical research on early imitation and joint attention which suggests that these approaches might be helpful in detecting delays in some specific cognitive areas.

There has been a dramatic increase in research on early imitation over the last two decades (Poulson, Nunes, & Warren, 1989). Recently, there has been a particular interest in how imitation informs models of early cognitive development (Meltzoff, Kuhl, & Moore, 1991; Meltzoff & Moore, 1995). The capacity to recall events which are perceptually unavailable, as shown in deferred imitation, is treated as a marker of a representational capacity (Mandler, 1988). The deferred imitation paradigm requires imitation of the model following a delay, in other words, from memory. Contrary to the classical Piagetian view of deferred imitation, which is hypothesized to emerge along with other symbolic skills toward the end of the second year, current empirical evidence points toward the end of the first year as the period where deferred imitation is first observed (Meltzoff, 1988; Bauer & Mandler, 1992). The deferred-imitation paradigm has recently been adapted for use with children with Down Syndrome (Meltzoff & Gopnik, 1993). The performance of children with Down Syndrome has been found to match the pattern found in normally-developing children very closely, although the older children in the sample (25–44 months) showed slightly stronger performances.

Joint attention is a social–cognitive phenomenon which requires that two individuals *know* that they are attending to the same object or event. This ability is usually observed toward the last quarter of the first year. Prior to 9 months of age, only simultaneous looking and gaze following are observed (Bakeman & Adamson, 1984; Butterworth & Jarrett, 1991). Researchers have begun to explore the development of children's joint attentional skills as "precursors" to their theories of mind and language skills (Moore & Dunham, 1995; Wellman, 1993). For instance, it has been demonstrated that adults' sensitivity to children's focus of attention plays an important role in infants' success at word learning (Harris, Jones, Brookes, & Grant, 1986). Mothers of children learning language at a normal rate made more references to objects that were the focus of the children's attention than mothers of slower language learners. In addition, a larger vocabulary has been found to be related to the amount of time spent in joint attentional focus (Tomasello, Mannle, & Kruger, 1986).

In contrast with the data on imitation from children with Down Syndrome suggesting that their imitation skills are similar to those of normal children, there is accumulating evidence for imitation and joint attention impairments in autistic children (Rogers & Pennington, 1991; Smith & Bryson, 1994). Since autism is not commonly diagnosed until at least the age of 3, it is extremely important to explore the potential application of the imitation paradigm for an earlier diagnosis of autism. This would require broad-based screening and later follow-up testing. The first steps toward that goal have recently been undertaken through a new prospective screening instrument for autism in infancy, the Checklist for Autism in Toddlers (CHAT; Baron-Cohen, Allen, & Ginsberg, 1992). This instrument checks for the presence of pretend play and joint attention behaviours. The CHAT was found to have good predictive value in a recent study of a group of 18-month-old children at risk for autism (siblings of already diagnosed children with autism) whose performances were compared to a group of randomly selected toddlers of the same age (Baron-Cohen et al., 1992). Four children from the high-risk group lacked both pretend play and joint attention and all of them were later diagnosed as autistic at 30 months. Following these encouraging findings, the first epidemiological study to attempt early screening for autism was recently carried out on a population of 16,000 18-month-olds (Baron-Cohen et al., 1996). A sample of children from the three groups identified in that population (Autism Risk, Developmental Delay Risk, No Risk) were recently tested for their imitation skills at the age of 20 months (Charman et al., in press). The infants at risk for autism produced less imitation than the developmentally delayed comparison group. Since imitation can be documented in infants as young as 9 months, there appears to be a potential for imitation paradigms to play a role in the early detection of autism.

Obviously, however, the real potential of these procedures for the early diagnosis and intervention of autistic children remains to be determined. At first glance, the deferred imitation and joint attention paradigms seem to tap very specific abilities that might have limited relevance for the assessment of general cognitive skills. However, there is empirical evidence for the value of joint attention skills in infancy in predicting verbal skills in early childhood (Tomasello & Todd, 1983; Dunham, Dunham, & Curwin, 1993). Studies relating early imitation skills to other cognitive skills are, to our knowledge, lacking. One of the main reasons for the dearth of literature on the predictive value of these imitative skills is that researchers have not paid much attention to the individual variability within the samples tested. Rather, they have focused on general stages in the development of these skills. Increased emphasis on individual variability is a task that researchers studying cognitive development will hopefully undertake in the near future in order to bridge the somewhat artificial (and unfortunate) gap between basic and applied research in infant cognition.

Summary

While the aforementioned new assessment tools are a substantial advancement over the poor predictability and limitations of past ones, they are still in need of validation, and few intervention programs are yet available to treat infants with information processing delays. Furthermore, these tests do not place sufficient emphasis on the role of the caregiver in infants' cognitive development. New tasks designed to assess imitation and joint attention skills show promise as a means of helping to identify children with specific delays. Their potential was shown in recent studies on imitation and joint attention where performances in these areas was used as early indicators of risk for autism.

The development of tasks or paradigms that could identify delays in specific cognitive skills is in accord with a view of normal cognitive development that has recently been gaining increasing respectability. According to this view, many cognitive abilities are specialized to process specific types of information (Gardner, 1983; Hirschfeld & Gelman, 1994). As such, many intellectual delays could be conceptualized as delays in some specific mental domain, as suggested by a recent model of autism arguing that a specific deficit in social understanding is at play (Baron-Cohen, 1995). With more fine-tuned assessment instruments will come more fine-tuned intervention programs that will be better suited to the needs of specific populations.

EVALUATION AND SUMMARY

The period of infancy is the least protracted period in human development. However, its brevity is countered by the dramatic and remarkable growth to the organism that occur during this short period. The amount and speed of development occurring in the socioemotional and cognitive areas is unmatched by any other period in the human life span. Consequently, the infant is vulnerable to biological and environmental hazards along the way. These potential hazards make it imperative to maintain the infant on an optimal developmental course, and underscore the significance and salience of prevention and intervention programs.

According to Shonkoff and Meisels (1990), investing in the future of our children by endorsing the cooperation and alliance of research scholars, practitioners, and educators with those who design and implement policy initiatives, is the fundamental purpose of early childhood intervention. However, the task is not an easy one, and the history of the early childhood intervention field is replete with controversy and challenges to methodology, program effectiveness, logistical constraints, and even ethical barriers. For interventions directed at either biological or environmental risk, methodological problems have frequently compromised the ultimate clarity of the findings and their interpretations and implications. Thus, effectiveness of the interventions and their cost–benefits have been challenged (e.g., Simeonsson, Cooper, & Scheiner, 1982). At the same time, however, most reviews and meta-analyses have found minimal to moderate effects (or better) of early intervention, and have underscored potential moderating influences (e.g., Guralnick, 1991). Whether specific factors, such as age at commencement and family involvement, are critical to success, remains a divided issue. However, it is important to note that most intervention programs have targeted cognitive competence — increased efforts to emphasize and integrate aspects of social competence could prove valuable.

Successful and optimal intervention is enhanced with valid and comprehensive assessments to identify the needs. The examples of strategies, paradigms and new procedures for assessing socioemotional and cognitive development described in the present chapter are rich and offer promise for ultimately providing a comprehensive portrayal of functioning in normally and aberrantly developing children, whether individually or in combination. However, additional research is warranted to integrate various strategies into more comprehensive ones, to validate these procedures, and to establish norms against which individual functioning can be compared. Alternatives to traditional assessments are indicated for young children as well as aberrantly developing children, as they add to the information obtained, provide comprehensive detail, and permit information that traditional forms of assessment would not generate on the child's developing abilities (Cicchetti & Wagner, 1990). Applying nontraditional assessment protocols or implementing research procedures into clinical contexts can pose its own difficulties. Integrating alternative and comprehensive strategies into already existing standardized batteries might be a means of facilitating this process and this approach might lead to improved predictive validity and a more representative portrayal of the infant's strengths and weaknesses (e.g., Cicchetti & Wagner, 1990; Zelazo, 1988). Moreover, direct observational procedures that are showing promise require assessment of their clinical utility and psychometric validity before any steps can be taken to create standardized comprehensive batteries which include them.

Recognition of the interrelations among domains is necessary to develop comprehensive and sensitive assessment instruments and to inform plans for intervention, because problems or referral issues in one domain may carry over or affect other domains of the child's development, or have ramifications or potential implications for other areas of development. As such, a multi-domain, multi-contextual approach where information is collected in multiple settings and a number of response domains are measured is recommended (Cicchetti & Toth, 1987; Achenbach, 1982). Such multi-contextual, multi-domain assessments have the added advantage of facilitating increased understanding of the contributions of constitutional (e.g., temperament), cognitive, and socioemotional (e.g., parent–child relationships) factors (Cicchetti & Wagner, 1990). Consistent with the focus of our chapter, there appears to be an important interface between emotion and cognition. According to Cicchetti and Wagner (1990), in studying how emotions such as joy, fear and other emotions develop, we are faced with processes such as memory, expectation, intentionality, abilities to distinguish people, and other domains of cognitive development. There are also data to support the contention that emotions influence other developmental domains (e.g., Motti, Cicchetti, & Sroufe, 1983). For example, it has been argued that there are possible links between language development and behavior disturbance, e.g., children who show delays or disturbance in language are at risk for development of emotional and behavioral problems (e.g., Baker & Cantwell, 1987, cited in Mundy & Willoughby, 1996). "Indeed, continued research and theorizing about the relation between emotional development and other ontogenetic domains may well prove to be a rich prognostic index for later adaptation" (Cicchetti & Wagner, 1990, p. 264).

Increased research in the area of infant assessment and intervention can only advance our thinking and enhance knowledge, so that approaches and services to increase competence are made generally available. In this way, models of "well-being" or "competence," and factors which contribute to such models, would be elucidated. Such models have the advantage of modifying current conceptualizations of defining normality or competence through the absence of abnormality. We know, for example, that fostering physical and mental health and growth in babies is important, and that socioemotional and cognitive competence are, in part, aided by good nutrition, a healthy and stimulating environment, a positive parent–child relationship, and the provision of opportunities for structured and independent learning and self-regulation.

Prospective, longitudinal studies of no and at-risk community populations which seek to follow families from infancy (e.g., Werner & Smith, 1992) or across more than one generation over developmental transitions (e.g., Serbin, Cooperman, Peters, Lehoux, Stack, & Schwartzman, 1997; Fagot, Pears, Capaldi, Crosby, & Leve, 1997) offer us an important means of documenting factors which contribute to resilience and risk. The burgeoning field of developmental psychopathology provides us with others (e.g., Lewis &

Miller, 1990). These approaches will ultimately enlighten us about intervention and prevention strategies, and the components of models of competence. Efforts to evaluate factors and models in light of ethnic diversity is another important step, as what is defined as "competent" in one culture may not be so in another culture, and cultural expectations may vary as a function of age and stage.

Models based on theories of maternal and environmental deprivation have been replaced by "transactional" models of development (Sameroff & Chandler, 1975; Sameroff & Bartko, this volume), in recognition of the view that the child is an influential component of its own environment and socio-cultural milieu. As such, this approach has had the effect of modifying the focus of early intervention programs from either only the child or only the parent to a perspective which recognizes and includes the roles of both caregiver and child within a changing context (Black, 1991). The infant is embedded in a social environment and that environment may be optimal or non-optimal, adaptive or maladaptive. Moreover, the revolution in views of the infant has demonstrated that there are a complex series of variables impinging and influencing development. Thus, simple linear prediction models must be discarded.

The evolving and changing nature of early intervention and prevention is considered a positive direction (Shonkoff & Meisels, 1990) — it reflects the dynamic nature of development and the rapid increase in research that extends our knowledge and fosters and encourages continuation of the research quest. Because, "ultimately, early childhood intervention must reflect our best attempts to translate ever-growing knowledge about the process of human development into the formation of the best kind of environment in which a child can grow" (Shonkoff & Meisels, 1990, p. 27). While at least the *potential* of early intervention to reap long term benefits has been demonstrated, simple generalizations from successes to policy are not recommended; rather, long term effects are embedded in a wider context of family, community and school processes which require attention (Woodhead, 1988). In essence, the mechanisms underlying positive effects are not always clear and continued research efforts in this area are warranted. Finer tuned and valid assessment tools to identify problems, coupled with clear intervention strategies tightly linked to assessment and individually tailored to the child's needs, is a direction worthy of pursuit. Prevention is another.

In our view, there is a clear need for an improved research–clinical (scientist–practitioner) interface. This "…unfortunate schism that exists between professionals involved in research and those engaged in service provision" (Cicchetti & Wagner, 1990, p. 270), is not a new story. Indeed, an approach which bridges the gap between research and practice and integrates the apparent fragmentation of professional disciplines in the field has been advocated before (e.g., Minnes & Stack, 1990). Moreover, new implications and applications from current research paradigms, while promising and underscoring the increasing appreciation we have for the breadth of infant competencies, simultaneously bring with them new challenges and caution. In the words of Barr and Zelazo (1989), "it seems paradoxically that the challenges are at one and the same time both compelling and not compelling. They challenge clinicians to act differently, but do not invest the decision with certainty" (p. 9). Perspectives on which our assessment and intervention protocols are based cannot be accepted at face value (Wolff, 1989); rather, openness to new paradigms, accountability, scientific rigor, and continual evaluation and re-evaluation in light of new information are the makings of good science. Clinical practice is affected by our covert notions of development (Wolff, 1989) and as these conceptions of development change, implications for clinical practice come to the fore. A new science of prevention and a national prevention research agenda is one suggestion (Coie et al., 1993) which would encourage bringing together science and practice.

REFERENCES

Achenbach, T. M. (1982). *Developmental psychopathology* (2nd ed.). New York: John Wiley.

Ainsworth, M. D. S., Bell, S. M., & Stayton, D. J. (1974). Infant–mother attachment and social development: "Socialization" as a product of reciprocal responsiveness to signals. In M. P. M. Richards (Ed.), *The integration of a child into a social world*. London: Cambridge University Press.

Ainsworth, M. D. S., & Wittig, B. A. (1969). Attachment and exploratory behaviour of one year olds in a strange situation. In B. M. Foss (Ed.), *Determinants of Infant Behaviour* (Vol. 4). London: Methuen.

Ainsworth, M. H., Blehar, M. C., Waters, E., & Wall, S. (1978). *Patterns of attachment: A psychological study of the strange situation*. Hillsdale, NJ.: Erlbaum.

Alessandri, S. M., & Lewis, M. (1993). Parental evaluation and its relation to shame and pride in young children. *Sex Roles, 29,* (56).

Ames, E. W. (1997, January). *The development of Romanian orphanage children adopted to Canada*. Final report for National Welfare Grants Program Human Resources Development Canada. Burnaby, British Columbia: Simon Fraser Unviersity, Psychology Department.

Bakeman, R., & Adamson, L. (1984). Coordinating attention to people and objects in mother–infant interactions. *Child Development, 55,* 1278–1289.

Baron-Cohen, S. (1995). *Mindblindness: An essay on autism and theory of mind*. Cambridge, MA: MIT Press.

Baron-Cohen, S., Allen, J., & Ginsberg, C. (1992). Can autism be detected at 18 months? The needle, the haystack and the CHAT. *British Journal of Psychiatry, 161,* 839–842.

Baron-Cohen, S., Cox, A., Baird, G., Swettenham, J., Nightingale, N., Morgan, K., Drew, A., & Charman, T. (1996). Psychological markers in the detection of autism in infancy in a large population. *British Journal of Psychiatry, 168,* 158–163.

Barr, R. G., & Zelazo, P. R. (1989). Do challenges to developmental paradigms compel changes in practice? An introduction. In P. R. Zelazo & R. G. Barr (Eds.), *Challenges to developmental paradigms*. Hillsdale, NJ: Erlbaum.

Bauer, P. J., & Mandler, J. M. (1992). Putting the horse before the cart: The use of temporal order in recall of events by one-year old children. *Developmental Psychology, 28,* 441–452.

Bayley, N. (1969). *Bayley Scales of Infant Development*. New York: Psychological Corporation.

Bayley, N. (1993). *Bayley Scales of Infant Development*. San Antonio, TX: Psychological Corporation.

Benasich, A. A., & Bejar, I. I. (1992). The Fagan Test of Infant Intelligence: A critical review. *Journal of Applied Developmental Psychology, 13,* 153–171.

Benasich, A. A., Brooks-Gunn, J., & Clewell, B. C. (1992). How do mothers benefit from early intervention programs? *Journal of Applied Developmental Psychology, 13,* 311–362.

Biringen, Z., Robinson, J. L., & Emde, R. N. (1988). The emotional availability scales. Unpublished manuscript. University of Colorado Sciences Center, Denver.

Black, M. M. (1991). Early intervention services for infants and toddlers: A focus on families. *Journal of Clinical Child Psychology, 20* (1), 51–57.

Bornstein, M. H., & Lamb, M. E. (1992). *Development in infancy: An introduction* (3rd ed.). NY: McGraw-Hill.

Bornstein, M. H., & Sigman, M. D. (1986). Continuity in mental development from infancy. *Child Development, 57,* 251–274.

Brazelton, R. B. (1984). *Neonatal behavioral assessment scale*. Philadelphia: J. B. Lippincott.

Bretherton, I. (1987). New perspectives on attachment relations: Security, communication, and internal working models. In J. Osofsky (Ed.), *Handbook of infant development* (pp.1061–1100). New York: Wiley.

Butterfield, P. M. (1993). Responses to IFEEL pictures in mothers at risk for child maltreatment. In R. N. Emde, J. D. Osofsky, & P. M. Butterfield (Eds.), *The IFEEL pictures: A new instrument for interpreting emotions* (pp.161–173). Madison, CT: International University Press.

Butterfield, P. M. & Ridgeway, D. (1993). The IFEEL Pictures: Description, administration and sexicon. In R. N. Emde, J. D. Osofsky, & P. M. Butterfield (Eds.), *The IFEEL pictures: A new instrument for interpreting emotions*. Madison, CT: International Universities Press.

Butterworth, B., & Jarrett, N. (1991). What minds have in common is space: Spatial mechanisms serving joint visual attention in infancy. *British Journal of Developmental Psychology, 9,* 55–72.

Caspi, A., & Silva, P. A. (1995). Temperamental qualities of age three predict personality traits in young adulthood: Longitudinal evidence from a birth cohort. *Child Development, 66,* 486–498.

Charman, T., Swettenham, J., Baron-Cohen, S., Cox, A., Baird, G., & Drew, A. (in press). Infants with autism: An investigation of empathy, pretend play, joint attention and imitation. *Developmental Psychology*.

Cicchetti, D., & Toth, S. (1987). The application of a transactional risk model to intervention with multi-risk maltreating familes. *Zero to Three, I,* 1–8.

Cicchetti, D., & Wagner, S. (1990). Alternative assessment strategies for the evaluation of infants and toddlers: An organizational perspective. In S. J. Meisels & J. P. Shonkoff, (Eds.), *Handbook of Early Childhood Intervention.* New York: Cambridge University Press.

Clark, R. (1985). The parent–child early relational assessment. Unpublished document available from Roseanne Clark, Ph.D., Department of Psychiatry, University of Wisconsin Medical School, Madison, WI.

Clark, R., Paulson, A., & Conlin, S. (1993). Assessment of developmental status and parent–infant relationships: The therapeutic process of evaluation. In C. H. Zeanah (Ed.), *Handbook of Infant Mental Health* (pp. 191–209). New York: Guilford.

Cohn, J. F., Campbell, S. B., & Ross, S. (1991). Infant response to the still-face paradigm at 6 months predicts avoidant and secure attachment at 12 months. *Development and Psychopathology, 3,* 367–376.

Cohn, J. F., & Tronick, E. Z. (1983). Three-month-old infants' reaction to simulated maternal depression. *Child Development, 54,* 185–193.

Coie, D., Watt, N. F., West, S. G., Hawkins, J. D., Asarnow, J. R., Markman, H. J., Ramey, S. L., Shure, M. B., & Long, B. (1993). The science of prevention: A conceptual framework and some directions for a National Research Program. *American Psychologist, 48* (10), 1013–1022.

Colombo, J., & Fagen, J. W. (Eds.). (1990). *Individual differences in infancy: Reliability, stability, and prediction.* Hillsdale, NJ: Erlbaum.

Crittenden, P. M. (1981). Abusing, neglecting problematic and adequate dyads: Differentiating by patterns of interaction. *Merrill-Palmer Quarterly, 27,* 201–218.

Denham, S. A., Lydick, S., Mitchell-Copeland, J., & Sawyer, L. (1996). Socioemotional assessment for atypical infants and preschoolers. In M. Lewis & M. W. Sullivan (Eds.), *Emotional development in atypical children.* Mahwah, NJ: Erlbaum.

Dunham, P., Dunham, F., & Curwin, A. (1993). Joint attentional states and lexical acquisition at 18 months. *Developmental Psychology, 29,* 827–831.

Easterbrooks, M. A., Lyons-Ruth, K., Biesecker, G., & Carper, A. (1996). Infancy predictors of emotional availabilty in middle childhood: The role of attachment and maternal depression. The L. P. Lipsitt (Chair), *Emotional availability in mother–child dyads: Predictors across development and risk status.* Symposium presented at the International Conference on Infant Studies, Rhode Island, USA.

Emde, R. N. (1987). Infant mental health: Clinical dilemmas, the expansion of meaning and opportunities. In J. D. Osofsky (Ed.), *Handbook of infant development* (pp. 1297–1320). New York: Wiley.

Emde, R. N. (1993). The collaborative history of the IFEEL Pictures. In R. N. Emde, J. D. Osofsky, & P. M. Butterfield (Eds.), *The IFEEL pictures: A new instrument for interpreting emotions.* Madison, CT: International Universities Press.

Emde, R. N., Osofsky, J. D., & Butterfield, P. M. (1993). *The IFEEL pictures: A new instrument for interpreting emotions.* Madison, CT: International Universities Press.

Fagan, J. F., III, & Detterman, D. K. (1992). The Fagan Test of Infant Intelligence: A technical summary. *Journal of Applied Developmental Psychology, 13,* 173–193.

Fagan, J. F., & Shepherd, P. A. (1987). *Fagan test of infant intelligence: Training manual.* Cleveland, OH: Infantest Corporation.

Fagan, J. F., & Singer, L. T. (1983). Infant recognition memory as a measure of intelligence. In L. P. Lipsett (Ed.), *Advances in infancy research* (Vol. 2, pp. 31–78). Norwood, NJ: Ablex.

Fagot, B. I., Pears, K. C., Capaldi, D. M., Crosby, L., & Leve, C. S. (1997). Becoming an adolescent father: Precursors and parenting. *Developmental Psychology,* manuscript under review.

Field, T. (1980). Interactions of preterm and term infants with their lower-and middle-class teenage and adult mothers. In T. Field, S. Goldberg, D. Stern, & A. Sostek (Eds), *High-risk infants and children: Adults and peer interactions* (pp. 113–132). New York: Academic Press.

Field, T. (1987). Affective and interactive disturbances in infants. In J. D. Osofsky (Ed.), *Handbook of infant development* (pp. 972–1005). New York: Wiley.

Field, T. M., Vega-Lahr, N., Scafidi, F., & Goldstein, S. (1986). Effects of maternal unavailability on mother–infant interactions. *Infant Behavior and Development, 9,* 473–478.

Fishbein, H. D. (1976). *Evolution, development and children's learning.* Pacific Palisades, CA: Goodyear Publishing.

Fogel, A. (1997). *Infancy: Infant, family, and society* (3rd ed.). New York: West.

Fogel, A. (1993). *Developing through relationships: Origins of communication, self, and culture.* Chicago: University of Chicago Press.

Gardner, H. (1983). *Frame of mind.* New York: Basic Books.

Greenspan, S. (1983). Parenting in infancy and early childhood: A developmental structuralist approach to delineating adaptive and maladaptive patterns. In J. Sasserath & R. Hoekelman (Eds.), *Minimizing high risk parenting* (pp. 79–86). Skillman, NJ: Johnson & Johnson Baby Products.

Greenspan, S., & Lieberman, A. F., (1989). Infants, mothers, and their interaction: A quantitative clinical approach to developmental assessment. In S. I. Greenspan & G. H. Pollock (Eds.), *The course of life, Vol. 1: Infancy* (pp. 503–560). Madison, CT: International Universities Press.

Guralnick, M. J. (1991). The next decade on the effectiveness of early intervention. *Exceptional Children, 58* (2), 174–183.

Harris, M., Jones, D., Brookes, S., & Grant, J. (1986). Relations between non-verbal context of maternal speech and rate of language development. *British Journal of Developmental Psychology, 4*, 261–268.

Hirschfeld, L. A., & Gelman, S. A. (Eds.) (1994). *Mapping the mind: Domain specificity in cognition and culture.* New York: Cambridge University Press.

Howes, C. (1987). Social competence with peers in young children: Developmental sequences. *Developmental Review, 7*, 252–272.

Isabella, R. A. (1993). Origins of attachment: Maternal interactive behavior across the first year. *Child Development, 64*, 605–621.

James, W. (1890). *Principles of psychology.* NY: Holt.

Kagan, J. (1994). *Galen's prophecy: Temperament in human nature.* New York, NY: Basic Books.

Kagan, J., Kearsley, P., & Zelazo, P. (1978). *Infancy: Its place in human development.* Cambridge, MA: Harvard University Press.

Karmiloff-Smith, A. (1996, Fall). The connectionist infant: Would Piaget turn in his grave? *Newsletter of the Society for Research in Child Development, 10*, 1–3.

Kaye, K. (1982). *The mental and social life of babies: How parents create persons.* Chicago, IL: The University of Chicago Press.

Klahr, D. (1993). Information-processing approaches to cognitive development. In M. H. Bornstein & M. E. Lamb (Eds.), *Developmental psychology* (3rd ed., pp. 273–335). Hillsdale, NJ: Erlbaum.

Kopp, C. B. (1987). Developmental risk: Historical reflections. In J. Osofsky (Ed.), *Handbook of infant development* (2nd ed., pp. 881–912). New York: Wiley.

Kopp, C. B., & McCall, R. B. (1982). Predicting later mental performance for normal, at-risk and handicapped infants. In P. B. Bates & O. G. Brim (Eds.), *Life-span development and behavior* (Vol. 4, pp. 33–61). San Diego, CA: Academic Press.

Lamb, M. E., & Bornstein, M. H. (1987). *Development in infancy* (2nd ed.). New York: Random House.

Lecuyer, R., & Streri, A. (1994). How should intelligence be characterized in the infant? In A. Vyt, H. Bloch, & M. H. Bornstein (Eds.), *Early child development in the French tradition: Contributions from current research* (pp. 75–90). Hillsdale, NJ: Lawrence Erlbaum.

Lewis, M. (1971). Individual differences in the measurement of early cognitive growth. In J. Hellmuth (Ed.), *Exceptional infant* (Vol. 2, pp. 172–210). Banbridge Island, WA: Brunner/Mazzel.

Lewis, M. (1987). Social development in infancy and early childhood. In J. D. Osofsky (Ed.), *Handbook of infant development* (pp. 419–493). New York: Wiley.

Lewis, M. (1992). *Shame, the exposed self* (pp.419–493). New York: Free Press.

Lewis, M. (1996, November). Emotional development and social competence. Invited paper presented at Improving Competence across the Lifespan Conference; Centre for Research in Human Development, Concordia University, Montreal, Canada.

Lewis, M., Alessandri, S. M., & Sullivan, M. W. (1992). Differences in shame and pride as a Function of Children's Gender and Task Difficulty. *Child Development, 63*, 630–638.

Lewis, M., & Michalson, L. (1983). *Children's emotions and moods: Developmental theory and measurement.* New York: Plenum.

Lewis, M., & Miller, S. M. (1990). *Handbook of developmental psychopathology.* New York: Plenum Press.

Mandler, J. M. (1988). How to build a baby: On the development of an accessible representational system. *Cognitive Development, 3*, 113–136.

McCall, R. B. (1979). The development of intellectual functioning in infancy and the prediction of later IQ. In J. D. Osofsky (Ed.), *Handbook of infant development* (pp. 707–741). New York: Wiley.

McCall, R. B. (1981). Early predictors of later IQ: The search continues. *Intelligence, 5*, 141–147.

McCall, R. B. (1994). What process mediates predictions of childhood IQ from infant habituation and recognition memory? Speculations on the roles of inhibition and rate of information processing. *Intelligence, 18*, 107

McCall, R. B., & Carriger, M. S. (1993). A meta-analysis of infant habituation and recognition memory performance as predictors of later IQ. *Child Development, 64*, 57–79.

Meltzoff, A. N. (1988). Infant imitation and memory: Nine-month-olds in immediate and deferred tests. *Child Development, 59*, 217–225.

Meltzoff, A., & Gopnik, A. (1993). The role of imitation in understanding persons and developing a theory of mind. In S. Baron-Cohen, H. Tager-Flusberg, & D. J. Cohen (Eds.), *Understanding other minds: Perspectives from autism* (pp. 335–366). Oxford, England: Oxford University Press.

Meltzoff, A. N., Kuhl, P. K., & Moore, M. K. (1991). Perception, representation, and the control of action in new-borns and young infants: Toward a new synthesis. In M. J. S. Weiss & P. R. Zelazo (Eds.), *Newborn attention: Biological constraints and the influence of experience* (pp. 377–411). Norwood, NJ: Ablex.

Meltzoff, A. N., & Moore, M. K. (1995). Infants' understanding of people and things: From body imitation to folk psychology. In J. L. Bermudez, A. Marcel, & N. Eilan (Eds.), *The body and the self* (pp. 43–69). Cambridge, MA: MIT Press.

Minnes, P. M., & Stack, D. M. (1990). Research and practice with congenial amputees: Making the whole greater than the sum of its parts. *International Journal of Rehabilitation Research, 13,* 151–160.

Montagu, A. (1986). *Touching: The human significance of the skin* (3rd ed.). New York: Harper & Row.

Moore, C., & Dunham, P. (Eds.) (1995). *Joint attention: Its origins and role in development.* Hillsdale, NJ: Erlbaum.

Motti, F., Cicchetti, D., & Sroufe, L. A. (1983). From infant affect expression to symbolic play: The coherence of development in Down Syndrome children. *Child Development, 54,* 1168–1175.

Mundy, P., Kasari, C., & Sigman, M. (1992). Nonverbal communication, affective sharing, and intersubjectivity. *Infant Behavior and Development, 15,* 377–381.

Mundy, P., & Willoughby, J. (1996). Nonverbal communication, joint attention, and early socioemotional develop-ment. In M. Lewis & M. W. Sullivan (Eds.), *Emotional development in atypical children* (pp. 65–87). Mahwah, NJ: Erlbaum.

Musick, J. S., Clark, R., & Cohlen, B. (1981). *The mother's project: A program for mentally ill mothers of young children. In Infants: Their social environments* (pp. 111–127). Washington, DC: National Association for the Education of Young Children.

Osofsky, J. D., & Culp, A. M. (1993). Perceptions of infant emotions in adolescent mothers. In R. N. Emde, J. D. Osofsky, & P. M. Butterfield (Eds.), *The IFEEL pictures: A new instrument for interpreting emotions.* Madison, CT: International Universities Press.

Piaget, J. (1952). *The origins of intelligence in children.* New York: International University Press.

Poulin-Dubois, D. (in press). Infants' distinction between animate and inanimate objects: The origins of naive psy-chology. In P. Rochat (Ed.), *Early social cognition.* Hillsdale, NJ: Lawrence Erlbaum.

Poulson, C. L., Nunes, L. R. P., & Warren, S. F. (1989). Imitation in infancy: A critical review. *Advances in Child Development and Behavior, 22,* 271–298.

Rausch, P. B. (1984). A tactile and kinesthetic stimulation program for premature infants. In C. C. Brown (Ed.), *The many facets of touch* (pp. 100–106). Pediatric Round Table Series No. 10. Skillman, NJ: Johnson and Johnson Baby Products.

Ricks, M. H. (1985). The social transmission of parental behavior: Attachment across generations. In I. Bretherton & E. Waters (Eds.), Growing points of attachment theory and research. *Monographs of the Society for Research in Child Development, 50* (1–2, Serial No. 209).

Rogers, S. J., & Pennington, B. F. (1991). A theoretical approach to the deficits in infantile autism. *Development and Psychopathology, 3,* 137–162.

Rose, S. A., Feldman, J. F., & Wallace, I. F. (1988). Individual differences in infant information processing: Reli-ability, stability, and prediction. *Child Development, 59,* 1177–1197.

Rose, S. A., & Ruff, H. A. (1987). Cross-modal abilities in human infants. In J. D. Osofsky (Ed.), *Handbook of infant development* (2nd ed., pp. 318–362). New York: Wiley.

Rutter, M. (1987). Continuities and discontinuities from infancy. In J. D. Osofsky (Ed.), *Handbook of infant development* (pp. 1256–1296). New York: Wiley.

Rutter, M. (1981). *Maternal deprivation reassessed* (2nd ed.). Harmondsworth, Middlesex, England: Penguin.

Sameroff, A. J. (1975). Transactional models in early relations. *Human Development, 18,* 65–79.

Sameroff, A. J., & Chandler, M. J. (1975). Reproductive risk and the continuum of caretaking casualty. In F. D. Horowitz (Ed.), *Review of child development research* (Vol. 4). Chicago, IL: University of Chicago Press.

Sameroff, A. J., & Fiese, B. H. (1990). Transactional regulation and early intervention. In S. J. Meisels & J. P. Shonkoff (Ed.), *Handbook of early childhood intervention.* New York: Cambridge University Press.

Scafidi, F. A., Field, T. M., Schanberg, S. M., Bauer, C. R., Vega-Lahr, N., Garcia, R., Power, J., Nystrom, G., & Kuhn, C. M. (1986). Effects of tactile/kinesthetic stimulation on the clinical course an sleep/wake behaviour of preterm neonates. *Infant Behaviour and Development, 9,* 91–105.

Segal, L. B., Oster, H., Cohen, M., Caspi, B., Myers, M., & Brown, D. (1995). Smiling and fussing in seven-month-old preterm and full-term black infants in the still-face situation. *Child Development, 66,* 1829–1843.

Serbin, L. A., Cooperman, J. M., Peters, P. L., Lehoux, P. M., Stack, D. M., & Schwartzman, A. E. (1997). Inter-gen-erational transfer of psychosocial risk in women with childhood histories of aggression and/or withdrawal: Results from the Concordia longitudinal project. *Developmental psychology,* manuscript under review.

Seifer, R., & Schiller, M. (1995). The role of parenting, sensitivity, infant temperment and dyadic interaction in attachment theory and assessment. In E. Waters, B. E. Vaughn, G. Posada, & K. Kondo-Ikemura (Eds.), *Caregiving, cultural, and cognitive perspectives on secure-base behavior and models: New growing points of attachment theory and research. Monographs of the society for research in child development: Vol. 60* (pp. 146–174). Chicago, IL: University of Chicago Press.

Shonkoff, J. P., & Meisels, P. C. (1990). Early childhood intervention: The evolution of a concept. In S. J. Meisels & J. P. Shonkoff (Eds.), *Handbook of Early Childhood Intervention.* New York: Cambridge University Press.

Simeonsson, R. J., Cooper, D. H., & Scheiner, A. P. (1982). A review and analysis of the effectiveness of early intervention programs. *Pediatrics, 69,* 635–641.

Smith, I. M., & Bryson, S. E. (1994). Imitation and action in autism: A critical review. *Psychological Bulletin, 116,* 259–273.

Spelke, E. S., Phillips, A., & Woodward, A. L. (1995). Infants' knowledge of object motion and human action. In D. Sperber, A. J. Premack, & D. Premack (Eds.), *Casual cognition: A multidisciplinary debate* (pp. 44–77). Oxford, England: Clarendon Press.

Sroufe, L. A. (1979). The coherence of individual development. *American Psychologist, 34,* 834–841.

Sroufe, L. A., & Wunsch, J. P. (1972). The development of laughter in the first year of life. *Child Development, 43,* 1326–1344.

Stack, D. M., & Muir, D.W. (1992). Adult tactile stimulation during face-to-face interaction modulates five-month-olds' affect and attention. *Child Development, 63,* 1509–1525.

Stack, D. M., & Muir, D. W. (1990). Tactile stimulation as a component of social interchange: New interpretations for the still-face effect. *British Journal of Developmental Psychology, 8,* 131–145.

Stern, D. N. (1985). *The Interpersonal World of the Infant.* New York, NY: Basic Books.

Szajnberg, M., & Skrinjaric, J. (1993). Perceptions of infant affect in mothers of prematures. In R. N. Emde, J. D. Osofsky, & P. M. Butterfield (Eds.), *The IFEEL Pictures Assessment.* Madison, CT: International Universities Press.

Thompson, R. A. (1993). Socioemotional Development: Enduring issues and new challenges. *Developmental Review, 13,* 372–402.

Tomasello, M., Kruger, A. C., & Ratner, H. H. (1993). Cultural learning. *Behavioral and Brain Sciences, 16,* 495–552.

Tomasello, M., Mannle, S., & Kruger, A. (1986). The linguistic environment of one to two year old twins. *Developmental Psychology, 22,* 169–176.

Tomasello, M., & Todd, J. (1983). Joint attention and lexical acquisition style. *First Language, 4,* 197–212.

Tronick, E. Z., Als, H., Adamson, L., Wise, S., & Brazelton, T. B. (1978). The infant's response to entrapment between contradictory messages in face-to-face interaction. *Journal of the American Academy of Child Psychiatry, 17,* 1–13.

Vernon, P. E. (1980). *Intelligence: Heredity and environment.* San Francisco: W. H. Freeman.

Waters, E., & Sroufe, L. A. (1983). Social competence as a developmental construct. *Developmental Review, 3,* 79–97.

Wellman, H. M. (1993). Early understanding of mind: The normal case. In S. Baron-Cohen, H. Tager-Flusberg, D. Cohen, & F. Volkmar (Eds.), *Understanding other minds: Perspectives from autism* (pp. 10–39). Oxford, England: Oxford University Press.

Werner, E. E., & Smith, R. S. (1992). *Overcoming the odds: High risk children from birth to adulthood.* Ithaca, NY: Cornell University Press.

Wolff, P. (1989). The concept of development: How does it constrain assessment and therapy? In P. R. Zelazo & R. G. Barr (Eds.), *Challenges to developmental paradigms: Implications for theory assessment and treatment.* Hillsdale, NJ: Erlbaum.

Woodhead, M. (1988). The psychology informs public policy: The case of early childhood intervention. *American Psychologist, 43* (6), 443–454.

Zahn-Waxler, C., & Wagner, E. (1993). Caregivers' interpretations of infant emotions: A comparison of depressed and well mothers. In R. N. Emde, J. D. Osofsky, & P.M. Butterfield (Eds.), *The IFEEL pictures: A new instrument for interpreting emotions.* Madison, CT: International Universities Press.

Zelazo, P. R. (1979). Reactivity to perceptual–cognitive events: Application for infant assessment. In R. B. Kearsley & I. E. Sigel (Eds.), *Infants at risk: Assessment of cognitive functioning* (pp. 49–83). Hillsdale, NJ: Erlbaum.

Zelazo, P. R. (1988). An information processing paradigm for infant–toddler mental assessment. In P. M. Vietze & H. G. Vaughan, Jr. (Eds.), *Early identification of infants with developmental disabilities* (pp. 299–317). Philadelphia: Grune and Stratton.

Zelazo, P. R., & Stack, D. M. (1997). Attention and information processing in infants with Down Syndrome. In J. A. Burack & J. T. Enns (Eds.), *Attention, development and psychopathology* (pp. 123–146). NY: Guilford Press.

4

THE SOCIALIZATION OF SOCIOEMOTIONAL COMPETENCE*

Nancy Eisenberg

Arizona State University

The purpose of this chapter is to summarize work on the socialization of socioemotional competence. I begin my review by laying out my conceptual framework and essential definitions; then I discuss some of the findings in the literature on the socialization of emotion, including empathy-related responding. Although there is little research on the socialization of socioemotional competence in families undergoing stress, I briefly consider this topic in closing.

INITIAL THOUGHTS ON THE SOCIALIZATION OF SOCIOEMOTIONAL COMPETENCE

A basic assumption in my work is that socioemotional competence is based, at least in part, on individuals' abilities to regulate their emotion and emotionally driven behavior. People who are unable to modulate the intensity and duration of their internal emotional responses and emotionally driven behavior are likely to be physiologically overaroused and to behave in ways that do not foster constructive social interactions (Eisenberg & Fabes, 1992). In addition, they are unlikely to focus on and learn about issues of importance in emotionally evocative contexts (Hoffman, 1983). Typical reactions to emotional over-arousal may vary across individuals: some people may respond with highly inhibited or overcontrolled behavior whereas others may be undercontrolled or out-of-control (e.g., Block & Block, 1980; Kagan, in press; Robins, John, Caspi, Moffitt, & Stouthamer-Loeber, 1996). In either case, social and emotional competence are likely to be compromised. Thus, it is useful to focus on individual differences in emotionality and regulation, and on factors that may enhance or undermine critical regulatory processes. Of course, other processes and

* Writing of this chapter and the research described were supported by grants from the National Science Foundation (DBS-9208375) and the National Institutes of Mental Health (1 R01 HH55052) and Research Scientist Development and Research Scientist Awards from the National Institute of Mental Health (K02 MH00903 and K05 M801321) to Nancy Eisenberg.

Improving Competence across the Lifespan, edited by Pushkar *et al.*
Plenum Press, New York, 1998.

mechanisms such as individual differences in children's sociocognitive capacities also are relevant to social and emotional functioning, albeit not discussed in this chapter.

Eisenberg and Fabes (1992; Eisenberg, Fabes, & Guthrie, in press) proposed that dispositional differences in frequency and intensity of individuals' emotional reactions and in people's abilities to regulate themselves are linked in systematic ways to a host of components of social and emotional functioning. Consistent with this argument, relatively low levels of regulation generally have been associated with uncontrolled, nonconstructive social behavior, reactive aggression, low prosocial behavior, susceptibility to negative affectivity, and social rejection (Block & Block, 1980; Caspi, Henry, McGee, Moffitt, & Silva, 1995; Coie, Dodge, & Kupersmidt, 1990; Eisenberg et al., 1995; Olweus, 1980; Pulkkinen, 1982). In contrast, socially competent behavior, including prosocial behavior, sympathy, popularity, and socially appropriate behavior, has been linked to relatively high levels of regulation in school children or college students and low levels of negative emotionality (Eisenberg et al., 1993, 1995, 1997; Eisenberg, Fabes, Karbon, et al., 1996; Eisenberg, Fabes, Murphy, et al., 1996). Moreover, regulation is more important for predicting low levels of problem behavior if children are prone to intense and frequent negative emotions (Eisenberg, Fabes, Guthrie, et al., 1996).

My goal is not to review this literature in detail, but to make the point that individual differences in children's emotionality and regulation do appear to be relevant for understanding their social and emotional competence. If this is the case, then one reasonable way to think about the socialization of socioemotional competence is to consider how various parental practices and characteristics may affect children's emotional reactions to evocative or stressful situations and/or children's abilities to regulate their internal emotional states, their emotionally driven behavior, and situations eliciting emotion and stress.

The degree to which individuals typically respond intensely to emotional events and stimuli appears to have a temperamental basis (Larsen & Diener, 1987; Plomin & Stocker, 1989); thus, parental practices may have only a modest to moderate effect on individual differences in experienced intensity of children's emotional reactions. However, as is discussed shortly, socializers' behaviors and actions may influence how children interpret specific situations and whether (and to what degree) evocative situations come to elicit emotion. Moreover, although it appears that many aspects of emotional and behavioral regulation also have a temperamental basis (see Rothbart & Bates, in press), children can learn to regulate their behavior to some degree and to cope with stressful situations. Thus, it is likely that socializers have a substantial effect on children's socioemotional competence by influencing how and when children regulate their behavior.

DEFINITIONAL ISSUES

A focus in our work has been on empathy-related emotional reactions. Thus, it is essential that empathy-related responses are differentiated prior to reviewing the empirical findings.

Empathy frequently has been defined as an emotional response stemming from the recognition of another's emotional state or condition — a response that is very similar or identical to what the other person is feeling or might be expected to feel in the situation. *Sympathy* often may be a consequence of empathizing; it consists of feelings of concern or sorrow for another based on the recognition of another's emotional state or situation (Eisenberg, Fabes, Murphy, et al., 1996). Sympathy may not always stem from empathy; sometimes it may be the outcome of solely cognitive processes such as putting oneself in another's role or assessing information from memory relevant to the other person's situation.

Although empathy often may result in sympathy, it also may produce a self-focused, egoistic reaction (e.g., anxiety, discomfort) labeled *personal distress* (Batson, 1991). This distinction is critical because sympathy appears to be associated with an orientation toward the other person's needs and prosocial (particularly altruistic) behavior whereas people who experience personal distress tend to be self-focused and assist others primarily when that is the easiest way to make themselves feel better (Batson, 1991; Eisenberg & Fabes, 1990). Moreover, sympathy is related to social competence in children (Eisenberg, Fabes, Murphy, et al., 1996) whereas personal distress may be associated with relatively low social competence (see Eisenberg & Fabes, 1992; Eisenberg et al., 1990).

Eisenberg and Fabes (Eisenberg, Fabes, Murphy, et al., 1994) have argued that personal distress entails empathic overarousal whereas sympathy is experienced if one's empathic emotion is optimally modulated. Consistent with this notion, children and adults exhibit higher skin conductance in response to distressing than sympathy-inducing situations (Eisenberg, Fabes, Schaller, Carlo, & Miller, 1991; Eisenberg, Fabes, Schaller, Miller, et al., 1991). Moreover, Eisenberg and Fabes have obtained initial evidence for the assertion that people prone to dispositional sympathy are well regulated, emotionally and behaviorally (Eisenberg, Fabes, Murphy, et al., 1996; Eisenberg & Okun, 1996) whereas those prone to personal distress are low in regulation and prone to intense negative emotion (Eisenberg et al., 1994; Eisenberg & Okun, 1996). Thus, examination of the socialization of sympathy and personal distress is a useful vehicle for learning about the socialization of emotion regulation and socioemotional competence, and I refer to work on this topic frequently.

SOCIALIZATION AND SOCIOEMOTIONAL COMPETENCE

There are a number of ways that socializers may influence how children respond to, or cope with, emotionally evocative situations. For pragmatic purposes, it is more useful to divide socialization influences into the following overlapping categories: (a) parental reactions to children's emotions, (b) parental behaviors that influence children's focus on the self versus others; and (c) socializers' discussion of emotion.

Parental Reactions to Children's Emotions

Socializers' emotion-related reactions and practices would be expected to affect both the degree to which children express their emotions and their ability to regulate their emotion and emotionally driven behavior in appropriate ways. Through their reactions to their children's emotions, parents may teach their children how and when to express emotion, how to interpret others' emotional displays and behaviors, and ways to manage their emotion. However, this socialization process likely is bi-directional.

For example, how socializers react to the expression of emotion in general may affect the likelihood of children acknowledging and exhibiting negative emotions (Eisenberg et al., 1988) and whether situations involving these emotions become intrinsically distressing. Buck (1984) suggested that sanctions for emotional expressiveness are associated with physiological but not external markers of emotional responding in adults. This is because children who receive negative reactions to their displays of emotion gradually learn to hide their emotions but feel anxious when in emotionally evocative situations due to prior repeated associations between punishment and emotional expressivity. Analogously, Tomkins (1963) suggested that children learn to express distress without shame and to respond sympathetically to others if their parents respond openly with sympathy and nurturance to children's feelings of distress and helplessness.

Consistent with Buck's theorizing, Eisenberg et al. (1988) found that parental leniency with regard to the expression of emotion was positively related to dispositional empathy whereas parental restrictiveness was associated with facial personal distress when children viewed an empathy-inducing film. Eisenberg, Fabes, Schaller, Carlo, and Miller (1991) found that mothers' reports of restrictiveness in response to children's expression of their own negative emotion (e.g., sadness and anxiety) were associated with elementary school boys' tendencies to experience personal distress rather than sympathy. Boys exposed to negative reactions to their negative emotion seemed prone to experience distress when confronted with others' distress, but did not want others to know what they were feeling. Boys whose parents stress control of negative emotions such as anxiety and sadness probably have difficulty dealing with these emotions in social interactions.

Parental restrictiveness in regard to the expression of emotion may vary as a function of the emotion and the child's age. As just discussed, parental restrictiveness in regard to children's expression of their own directly experienced negative emotions, particularly ones that are inner directed and do not harm others (e.g., sadness rather than anger), may foster nonconstructive responses to negative emotion (particularly for boys). However, in other situations, parental restrictiveness may enhance children's awareness of when it is appropriate to express various emotions. Consistent with this line of reasoning, Eisenberg, Fabes, Schaller, Miller, et al. (1991) found that same-sex parents' restrictiveness in regard to emotional displays that could be *hurtful to others* (e.g., staring at a disfigured person) was associated with elementary school children's reports of situational and dispositional sympathy. Parents who discourage their children from expressing hurtful emotions may be educating their children about the effects of children's emotional displays on others. However, in another study, maternal restrictiveness in regard to the display of hurtful emotions was associated with kindergarten girls' (but not older children's or boys') physiological and facial arousal indicative of personal distress in a sympathy- inducing context. Supplemental analyses indicated that mothers who were restrictive in this regard with kindergarten girls were less supportive in general; thus, for younger children, such maternal restrictiveness may have reflected age-inappropriate restrictiveness or low levels of support (Eisenberg, Fabes, Carlo, Troyer et al., 1992), which might have fostered a tendency toward experiencing distress.

In general, parental practices that encourage young children's expression of emotion are associated with positive outcomes. Encouragement of the expression of emotion has been correlated with preschoolers' understanding of emotions (e.g., Denham, Zoller, & Couchoud, 1994) and parental positive reinforcement of children's emotions has been associated with teachers' ratings of social competence (Denham, Mitchell-Copeland, Standberg, Auerbach, & Blair, in press). Gottman, Katz, and Hooven (1996) found that parents who were aware and supportive of the expression of emotion had children who were physiologically well regulated and, therefore, could regulate their emotional arousal in social contexts; moreover, children's regulation predicted their competence with peers. In addition, parents with this perspective were less likely to derogate their children, and low derogation predicted high levels of children's competence with peers.

There may be an optimal level of parental encouragement of the expression of emotion. Roberts and Strayer (1987) found that a moderate level of parental encouragement of negative affect was linked to preschoolers' social competence. Nonetheless, the decline in children's social competence at high levels of parental encouragement was small.

However, not all findings are consistent. Denham (1993) found that maternal tenderness to two-year-old children's sadness, a supportive reaction to children's experience of emotion, was related to low levels of children's positive affiliation. Perhaps calming a sad child is effective in reducing the child's distress, but does not capitalize on opportunities for the child to learn to cope effectively. Alternatively, mothers may be particularly likely to calm children who tend to be sad or distressed, and such children would be expected to be deficient in social skills (Strauss, 1988).

By means of their reactions to children's emotions, socializers may teach children specific ways of dealing with emotions, as well as affect children's emotional responses. This teaching may be intentional or incidental to other parental goals. As just discussed, some parents emphasize hiding or controlling their emotions; others suggest or encourage techniques such as seeking social support. Still other parental reactions in evocative situations are problem-focused rather than designed to deal directly with children's emotion. Roberts and Strayer (1987) found that parental modeling of instrumental problem-solving responses when their children were upset (e.g., putting a bandage on the child when he/she was hurt) were related to children's social competence. Furthermore, parental emphasis on children's instrumental problem solving has been associated with sons' sympathy (Eisenberg, Fabes, Schaller, Carlo, & Miller, 1991). Thus, parental modeling and encouragement of problem-focused coping may foster children's emotional and social competence.

In recent research in our laboratory, we have studied the relation of quality of children's social functioning to six types of parental reactions to children's negative emotions (Eisenberg & Fabes, 1994; Eisenberg, Fabes, Carlo, & Karbon, 1992; Eisenberg, Fabes, & Murphy, 1996). Three reactions reflect relatively nonsupportive parental actions: minimizing of the child's negative emotion, punitive reactions, and parental distress reactions (e.g., discomfort) in response to the child's negative emotion. Three additional responses represent supportive reactions: encouragement of expressing emotion, emotion-focused reactions (i.e., comforting), and problem-focused reactions in which the child is encouraged or helped to deal with the problem. In an initial pilot study with 3- to 5-year-olds, Eisenberg, Fabes, Carlo, and Karbon (1992) found that parents' reports of minimizing reactions were associated with high frequencies of children's observed anger and low social competence; parental punitive reactions were associated with children's avoidant coping and inappropriate coping with anger (e.g., seeking revenge); and parental distress reactions also were correlated with children's inappropriate coping. In contrast, parental problem- and emotion-focused reactions and encouragement of expressing emotion tended to be positively associated with children's sociometric status or adults' reports of children's social skills and appropriate coping. However, due to the small sample size, many of these correlations were only marginally significant. In fact, in another study of 4- to 6-year-olds and their real-life anger reactions (Eisenberg & Fabes, 1994), there was relatively little support for the prediction that supportive parental reactions were positively related to socially competent functioning whereas nonsupportive, punitive practices are negatively related.

Relations between parental reactions to children's negative emotions and children's social functioning may become clearer with age. In a recent study (Eisenberg, Fabes, & Murphy, 1996), mothers' reports of responding to their third and sixth grade children's negative emotion with minimizing and/or punitive reactions were significantly correlated with maternal reports of children using avoidant coping and at least marginally correlated with low constructive coping and popularity, as well as teacher-reported popularity, social skills, and constructive coping. In contrast, maternal reports of problem-focused reactions were positively related to maternal reports of positive social functioning (e.g., constructive coping and social skills for the combined sample and popularity for girls), children's self-reported socially appropriate behavior, and teachers' reports of constructive coping and girls' popularity. Maternal encouragement of the expression of emotion was positively related to mothers' reports of constructive coping and children's self-reported social skills. Many, albeit not all, of these correlations remained significant when maternal reports of the intensity of children's negative emotion were controlled.

Mothers' reactions also were significantly correlated with children's quantity or quality of comforting of a crying infant. For girls, maternal punitive reactions were negatively related to the quality of comforting behavior whereas a moderate level of maternal encouragement of girls' expression of emotion was associated with high quantity and quality of comforting. For boys, maternal encouragement of expressing emotion was related to quantity and quality of comforting; problem-focused reactions were positively related to quantity of comforting; and emotion-focused reactions were positively related to quality of comforting. Thus, the findings suggest that supportive and problem-focused maternal reactions may foster boys' sensitivity to others' negative emotions and related comforting reactions, whereas, for girls, moderate maternal encouragement of expressivity and low levels of punitive reactions may be optimal. Perhaps, due to cultural expectations in regard to the expression and experience of emotion (Shields, 1995), boys need encouragement to experience their own and others' emotions if they are to be responsive to others. In contrast, girls who are encouraged to express moderate but not high levels of emotion may be likely to regulate their emotion and avoid overarousal due to high emotion and, hence, may experience sympathy. Recall that regulation of emotion has been linked to sympathy rather than distressed reactions to others' emotion (Eisenberg et al., 1994; Eisenberg & Okun, 1996).

In contrast to the mother-report data, there were few relations between fathers' reports of their reactions to children's negative emotions and their offsprings' social functioning. Paternal minimizing was negatively related to mothers' reports of children's popularity and teachers' reports of social skills and constructive coping. Unexpectedly, problem-focused reactions were negatively related to mothers' reports of girls' popularity and teachers' reports of girls' social skills. Fathers seemed to provide less problem-focused support if girls were viewed as socially competent by other adults. In addition, there was a significant quadratic relation between fathers' encouragement of the expression of emotion and quality of girls' comforting behavior: low and moderate levels of paternal encouragement for expression of emotion were associated with higher quality comforting among girls than high levels of paternal encouragement.

Overall, then, the findings provide some support for the view that maternal emotion-related reactions that are likely to reduce children's negative arousal and encourage a moderate degree of emotional expressivity foster positive social functioning. Of course, the data are correlational so it is impossible to determine cause and effect. Moreover, because the parent data are self-report, they may reflect parents' attitudes as much as their actual behavior. Nonetheless, the findings are consistent with the proposition that mothers enhance children's socioemotional development if they help them to regulate negative emotion and to learn instrumental ways of coping with stressors.

It is unclear why a similar pattern of findings was not obtained for fathers. Consistent with gender stereotypes (Shields, 1995), it is quite possible that emotion socialization is viewed as more appropriate for mothers than fathers. In this study (Eisenberg, Fabes, & Murphy, 1996), mothers scored high on supportive reactions to their child's negative emotion whereas fathers scored higher on punitive and minimizing reactions; similar findings recently were obtained by Zeman and Shipman (1996). Many fathers may attempt to socialize their child's emotional responding only if they perceive a problem. Research with observational methods is needed to determine whether there is a genuine difference in mothers' and fathers' emotion-related socialization behavior or whether gender stereotypes affect parents' and children's perceptions of such behavior.

Differences in fathers' and mothers' typical styles of parenting also may account for the fact that mothers', but usually not fathers', practices were associated with children's social functioning. Grusec and Goodnow (1994) argued that the impact of a particular disciplinary action is mediated by the child's judgment of its acceptability and appropriateness, which is based in part on parents' past style and perhaps culturally determined expectations of what is suitable behavior for fathers and mothers. For example, they suggested that the greater acceptability of fathers' power assertive techniques accounts for the fact that fathers' use of inductions (reasoning) and communicativeness do not predict moral internalization. If fathers are expected by children to respond to children's negative emotion with punitive and minimizing rather than supportive responses, their punitive/minimizing reactions have relatively little effect. Further, their supportive reactions may be viewed differently from those of mothers, which might alter their effects. For example, Bryant (1987) reported that paternal, but not maternal, nonindulgence and limit setting were related to enhanced empathic responding of boys and girls, both contemporaneously and longitudinally.

Parental Expressivity and Children's Expressivity and Socioemotional Competence

Parents' expression of emotion probably influences children's social and emotional competence by means of multiple mechanisms. First, it may influence children's social competence indirectly through its effects on children's own emotional expressivity. Children's expression of positive and negative emotions has been found to affect how children are perceived and whether they are liked by peers; for example, preschool children who express positive emotion are liked by peers (Sroufe, Schork, Motti, Lawroski, & LaFreniere, 1984). Second, parental expressiveness may influence children's abilities to interpret and understand others' emotional reactions, which could influence children's social competence. Third, parental expression of emotion (e.g., hostility toward the child, anger) may influence children's socioemotional competence relatively directly through mechanisms such as shaping children's feelings about themselves and others. This third method of influence is discussed in a later section on the quality of the parent–child relationship.

The Link between Parental and Child Expressivity. The degree to which parents exhibit emotion appears to be related to children's expression and reports of their own emotions. Balswick and Avertt (1977) and Halberstadt (1986) found that college students who rated their families as relatively expressive tended to be more expressive themselves than were students who rated their families as less expressive. Similarly, Denham (1989) found that mothers who displayed positive emotions had toddlers who tended to display positive rather than negative emotion. In addition, mothers who displayed more anger had toddlers who tended to display negative emotional expressions and relatively low levels of positive emotion. Preschoolers whose mothers reported intense emotions (positive or negative) also displayed more emotions at preschool; children of mothers who reported higher levels of externalizing emotion were lower on an index of positive versus angry emotion (Denham & Grout, 1993).

Moreover, it appears that parents' empathy-related reactions are correlated with those of their children. In several studies we obtained parents' reports of their own tendencies to respond with sympathy and personal distress, and also assessed children's empathic tendencies in the laboratory. Generally we have found that sympathetic parents tend to have same-sex children who are sympathetic and/or unlikely to experience personal distress when exposed to others in need (Eisenberg, Fabes, Schaller, Carlo, & Miller, 1991; Eisenberg, Fabes, Carlo, Troyer, et al., 1992; Fabes, Eisenberg, & Miller, 1990). In addition, mothers' dispositional personal distress has been linked with both sons' and daughters' inappropriate positive emotion in response to viewing others in distress (Fabes et al., 1990) and with either markers of girls' personal distress (Eisenberg, Fabes, Carlo, Troyer et al., 1992) or low levels of girls' vicarious emotional responding (Fabes et al., 1990). In another study in which mothers and children watched an empathy-inducing film together, mothers who tended to exhibit facial distress and heart-rate acceleration during the film had children who did likewise (Eisenberg, Fabes, Carlo, Troyer, et al., 1992).

The mechanisms that account for similarities in parents' and children's expressiveness are not clear. Modeling of parents' expressive style is one possible explanation; genetic factors also may account for the parent–child similarities. In addition, expressive parents may reinforce the expression of emotion due to their philosophy about emotions (see Gottman et al., 1996).

The Link between Family Expressiveness and Children's Emotional Understanding and Social Functioning. It is not clear that familial expressiveness influences children's understanding of emotion in a consistent manner (Cassidy, Parke, Butkovsky, & Braungart, 1992). In fact, in one study, adults from expressive families were less able to decode unknown others' emotions than were their peers (Halberstadt, 1986). In another study, maternal expression of anger in the home was related to a low level of understanding of anger, but not of other emotions (Garner, Jones, & Miner, 1994). However, family emotional expressiveness has been positively associated with children's socioemotional competencies. For example, intensity of parents' positive expressiveness has been linked to high sociometric status and children's prosocial behavior (Boyum & Parke, 1995). Frequency of mothers' positive expressivity also has been correlated with preschoolers' caregiving behavior with a young sibling (Garner et al., 1994). Denham and Grout (1992) found children's social competence (assessed by teachers) was associated with mothers' reports of their own intense happiness and relatively infrequent tension.

Frequency of familial or parental expression of general emotionality (negative as well as positive emotion) also has been positively related to children's social status (Bronstein, Fitzgerald, Briones, Pieniadz, & D'Ari, 1993; Cassidy et al., 1992) and, to some degree, with socially appropriate behavior (for boys) and low levels of psychological problems among older children (Bronstein et al., 1993). Denham and Grout (1992) found that frequent but low level expression of anger and anger caused by factors other than disobedience was related to social competence whereas frequency of anger was negatively related to children's prosocial responses.

Expression of negative emotion in the family seems to be positively related to social competence and prosocial behavior primarily when the expression of negative affect leads to discourse about feelings and explanations by parents for the negative affect (Denham & Grout, 1992, 1993; Dunn & Brown, 1994). Indeed, high overall levels of strong negative emotions such as anger by family members seem to be linked with low social competence in children, as indexed by low sociometric status (Boyum & Parke, 1995), low levels of prosocial behavior and sympathy (Denham, Renwick-DeBardi, & Hewes, 1994; Denham

& Grout, 1992; Eisenberg, Fabes, Carlo, Troyer, et al., 1992), less negotiation in conflict (Dunn & Brown, 1994), and poor performance on emotion-understanding tasks (Dunn & Brown, 1994). One critical factor may be whether the expression of externalizing emotions by parents is directed at the child or at others; Denham, Zoller, and Couchoud (1994) found that maternal negative emotions expressed during interaction with children were negatively correlated with emotional understanding. In contrast, maternal report of externalizing emotions which were not necessarily directed at the child was associated with children's understanding of emotions and prosocial behavior (Denham & Grout, 1993).

The expression of nonhostile softer negative emotions (e.g., loss, sadness) in the home also has been associated with daughters' vicarious emotional responding to others in distress (e.g., sympathy; Eisenberg, Fabes, Carlo, Troyer et al., 1992; Eisenberg, Fabes, Schaller, Miller et al., 1991), and social status (Boyum & Parke, 1995). For example, mothers' use of low level negative affect with sons was associated with social status and prosocial behavior (Boyum & Parke, 1995). Thus, low levels of negative emotion may serve an instructional purpose whereas high levels of intense negative emotions such as anger may be disruptive to learning about emotion.

Clarity of parental expressivity also may relate to children's social competence. Boyum and Parke (1995) found that parents' reports of clarity of expressiveness were associated with low levels of teacher-rated aggression, particularly for same-sex children. High clarity of mothers' negative expressiveness also predicted girls' prosocial behavior, whereas low clarity of positive expressiveness was associated with girls' verbal aggression. In a study of preschoolers, accuracy of mothers' anger simulations tended to be related to children's concern and sympathy in response to peers' emotional displays (Denham, Renwick-DeBardi, & Hewes, 1994). Thus, in general clarity of parental expressiveness, at least as reported or enacted by parents, has been associated with positive social functioning.

The Link between the Affective Quality of the Parent–Child Relationship and Children's Social Competence

There is considerable evidence in the attachment and peer relationship literatures that children who have affectively positive relationships with their parents are relatively high in social competence and low in problem behaviors (see Maccoby & Martin, 1983; Rubin, Bukowski, & Parker, in press; Thompson, in press). In addition, parental anger toward a child in difficult childrearing contexts has been linked to low levels of children's understanding of emotion (Denham et al., in press; Garner et al., 1994). Moreover, recently there has been discussion of the ways in which the quality of parent–child relationships may directly affect children's expression and regulation of their emotions. Specifically, attachment styles and relationships have been viewed as reflecting strategies for regulating emotion in interpersonal relationships (Bridges & Grolnick, 1995; Cassidy, 1994; Sroufe et al., 1984). For example, the securely attached infant whose parent is consistently and appropriately responsive to the infant's distress signals is viewed as learning that it is acceptable to exhibit distress and to actively seek out the assistance of others for comfort when stressed. In contrast, the avoidant infant, due in part to the parent's nonresponsiveness to his or her distress signals, appears to learn to inhibit emotional expressiveness as well as other-directed self-regulatory strategies (e.g., contact-seeking and maintaining behaviors; Bridges & Grolnick, 1995). Thus, attachment relationships are believed to influence children's socioemotional competence.

Consistent with the aforementioned hypotheses, mothers' tendencies to be appropriately responsive to the child's interactive behaviors have been associated with infants' increasing positive emotionality (e.g., Gianino & Tronick, 1988) and maternal support has been linked with the variety of coping strategies used by children to deal with stress, as well as the use of relatively appropriate strategies (e.g., avoidant strategies in uncontrollable situations; Hardy, Power, & Jaedicke, 1993). Furthermore, Kestenbaum, Farber, and Sroufe (1989) found that preschool children with secure attachments at 12 and 18 months of age were more empathic and more prosocial towards others. In contrast, anti-empathy responses (delight at others' discomfort, attempts to aggravate victims' condition, etc.) were observed to occur more often in children with anxious-avoidant attachment histories.

Nonsupportive, conflictual parent–child relationships also are related to children's expression of anger and hostility. For example, Matthews, Woodall, Kenyon, and Jacob (1996) found that frequency of negative behaviors exhibited by both parents and their early adolescent sons predicted sons' reports of hostility and expressing anger outwardly three years later after controlling for their initial levels of hostile attitudes and anger expression.

In our own work that was described previously, there also was some evidence that maternal supportive reactions to children's negative emotions were associated with higher levels of social competence in children. In contrast, maternal dismissing or punitive reactions were negatively related to some measures of social functioning. Recall, however, that there were few associations for fathers and some were contrary to prediction (Eisenberg, Fabes, & Murphy, 1996). In other studies, maternal warmth when telling empathy-inducing stories was correlated with second graders' (but not kindergartners') sympathy versus personal distress reactions and prosocial behavior (Fabes et al., 1994), and mothers' reports of comforting their children when the children experience negative emotion have been related to constructive anger reactions (Eisenberg & Fabes, 1994).

It is likely that the relation between quality of the parent–child relationships or interactions and children's socioemotional development is complex. As seems to be the case for the socialization of prosocial behavior (Eisenberg & Fabes, in press), supportive parent–child interactions may enhance social functioning only when combined with other socialization practices or influences that shape important relevant capacities. In the case of socioemotional development, those practices probably include ones that foster appropriate regulation of emotion. Moreover, the degree of support and warmth may need to be age-appropriate or it could undermine the process of children learning to regulate their own emotions.

Discussion of Emotion

The nature of socializers' communications about emotion and its regulation appears to contribute to children's socioemotional development. Emotion-related discussion in the family not only may communicate support, but also sharpen the child's awareness of emotional states and promote the development of an emotion-related conceptual system (Malatesta & Haviland, 1985). Moreover, discussions with parents contain information about self-related emotions such as guilt, shame, and pride, and provide connections between specific events or behaviors and emotional correlates or consequences.

Consistent with this view, socializers' willingness to discuss emotions with their children has been associated with children's awareness and understanding of others' emotional states (Denham, Cook, & Zoller, 1992). For example, references to feeling states by mothers and older siblings when children were 18 months of age have been associated with children's talking about feeling states at 24 months (Dunn, Bretherton, & Munn, 1987). A similar relation between mothers' emotion language and children's understanding of emotion was obtained with preschool children (Denham, Zoller, & Couchoud, 1994). In addition, Dunn, Brown, Slomkowski, Tesla, and Youngblade (1991) found that children's understanding of feelings at 40 months of age was associated with participation in family discourse about feelings and causality seven months earlier. With yet another sample, 3-year-olds who grew up in homes in which talking about feelings was frequent were better than peers at making judgments about others' emotions at age 6 (Dunn, Brown, & Beardsall, 1991).

Parental discussion of emotion has been linked not only with children's understanding of emotion, but also with their social competence in social interactions. For example, self-reported frequency of mothers' conversations about emotions has been correlated with preschoolers' social status with peers (Laird, Pettit, Mize, Brown, & Lindsey, 1994). However, Denham et al. (in press) found that frequency of parental discussion of emotion was marginally positively related to children's understanding of emotion but was negatively related to children's social competence.

In one of our own studies, mothers who mentioned their own sympathy and sadness to their child when watching an empathy-inducing film had sons who themselves reported more sympathy. However, maternal discussion of this sort was not related to children's facial reactions to the film. Mothers who verbally linked the events in the film to children's own experiences had children who exhibited more emotion to the film (sadness or distress for girls; concern or distress for boys) and sons who reported more sympathy or sadness (Eisenberg, Fabes, Carlo, Troyer et al., 1992).

It is likely that the type or function of parental discussion of emotion affects the relation between parental emotion-related discussion and children's social competence. As mentioned previously, we have obtained some evidence that fathers sometimes may try to guide and socialize children's emotional reactions more for children who are low than high in social competence (e.g., Eisenberg, Fabes, & Murphy, 1996). Similarly, Denham et al. (in press) found that parents' use of guiding and socializing language when discussing emotions with their preschoolers (e.g., "You really made me sad that time. I wish you wouldn't scream like that" or "Big kids don't cry so much") was correlated with low levels of children's positive reactions to peers' emotions and teacher-rated social competence, as well as children's internalizing emotion in the classroom. Frequency of discussion of emotion and parental attempts to explain or clarify the causes or consequences of feeling states generally were not related to children's social competence. Thus, a critical distinction may be whether parents use emotion language simply to clarify and teach about emotions or to try to correct children's inappropriate behavior.

Child Effects

It is likely that differences among children in temperament and other characteristics affect parental emotion-related socialization practices, parental expression of emotion and quanity, as well as quality of children's social and emotional responding. In a study with 4- to 6-year-olds, mothers reported that they were relatively punitive and avoidant in response to children's negative emotions if they viewed their children as high in negative emotionality and low in the ability to regulate attention. In contrast, mothers tended to report more supportive socialization practices if they believed their child to be attentionally regulated (Eisenberg & Fabes, 1994). In another study, fathers', as well as mothers', reports of reactions to their children's emotion were associated with their perceptions of their child's emotional intensity, with significant relations holding primarily for same-sex children (Eisenberg, Fabes, & Murphy, 1996). Of course, it is possible that parents' perceptions of their children reflected their own needs, values, or characteristics rather than reality; however, mothers' and fathers' perceptions of children's negative emotional intensity were positively intercorrelated so it is doubtful that parental perceptions were entirely inaccurate.

In another study (Fabes et al., 1994), mothers were asked to tell stories with emotionally evocative themes to their children from illustrated picture books with no words in them. Mothers' behaviors and affect in these story contexts varied as a function of their perceptions of their children's emotional reactivity and their children's age. Kindergartners were viewed by their mothers as more emotionally reactive to others' distress than were second graders, and mothers displayed more positive facial expressions when telling the emotionally evocative stories to kindergartners than second graders. It appeared that mothers of younger children attempted to minimize or modulate children's negative responses to the stories by maintaining a positive mood and avoiding too much negative facial emotion. This pattern of maternal emotional displays was especially true for younger children whose mothers believed that they were prone to relatively high emotional arousal when exposed to others' negative emotions. It is likely that mothers who viewed their children as especially vulnerable to distress or behavioral dysregulation as a consequence of experiencing vicariously induced negative emotion attempted to modulate the level of negative emotion experienced by their children.

In contrast, mothers who viewed their older children (i.e., second graders) as emotionally reactive to others' negative emotions tended to be less involved and warm in the storytelling context than were mothers who believed their children were less reactive. Thus, it seems that mothers who perceived their older children to be emotionally reactive may have "backed off" from socialization efforts that actively involved their children in a distressing experience.

In the same study, mothers' behaviors and emotion also were related to children's vicarious emotional responding and prosocial behavior. For younger children, mothers' use of positive versus negative facial expressions while telling the stories was positively related with children's helpfulness. In contrast, second graders' prosocial and emotional responses were predicted by maternal responsiveness (warmth and attempts to involve the child when telling the story) rather than expressiveness. Maternal responsiveness was correlated with relatively low levels of second graders' personal distress and relatively high sympathy and prosocial behavior. Thus, mothers varied their strategies for dealing with potentially evocative contexts depending on the age and perceived vulnerability of their child, and these variations predicted aspects of children's emotional and prosocial responding when the mother was not present.

Summary

It is clear that there are links between parental socialization practices and characteristics and the quality of children's emotional and social responding. However, it is also evident that some of the relations are complex and influenced by subtle variables. For example, the relation of parental expression of negative emotion with child socioemotional functioning may vary with the type of negative emotion expressed by parents, the appropriateness of the expression of that emotion, and whether the expression of externalizing negative emotions is occasional and accompanied by discussion of emotion or simply part of a hostile family environment. Similarly, links between parental discussion of emotion and children's socioemotional functioning may vary as a function of whether the emotion discussion is primarily for educational purposes (or simply part of a normal conversation) or reflects an attempt to rectify children's ongoing inappropriate behavior. Moreover, at this point in time, there are few data relevant to the issue of causality — that is, the degree to which parental reactions shape versus are in reaction to child characteristics and behavior. Most likely, causality is bidirectional and complex. Moreover, genetic factors may account in part for some associations between socializers' behaviors (e.g., expressivity) and children's emotionality or social behavior (e.g., children's affective balance or expressivity). In addition, other factors in the socialization context such as stress in the home or peer reactions to children's emotional displays undoubtedly play a role in the socialization of socioemotional competence. The issues of changes in the family environment and interventions in socioemotional responding are now considered briefly.

THE CHANGING FAMILY AND SOCIOEMOTIONAL DEVELOPMENT IN CHILDREN

Unfortunately, there is relatively little research directly pertaining to the effect of changes in family structure or economic circumstances on the socialization of socioemotional competence. Nonetheless, there is mounting evidence that level of familial stress can affect the quality of the parent–child relationship (Davies & Cummings, 1994), as well as parents' emotional state (e.g., degree of depression) and childrearing practices, with implications for children's socioemotional development (Conger et al., 1992, 1993; McLoyd, Jayaratne, Ceballo, & Borquez, 1994).

Conflict in the Family

Conflict in the family is not new to society and is present in varying degrees in many families. However, with the widespread occurrence and acceptance of divorce in many Western societies and the change of women's roles, it is likely that intense conflict is more common than at some times in the past, and often is relatively overt prior to and following divorce.

Familial conflict appears to account for some of the negative effects of divorce on children's functioning (Amato & Keith, 1991). Such conflict has been associated with behavior problems in children (Davies & Cummings, 1994). Resultant problem behaviors may be externalizing or internalizing problems; the former appear to be more likely when parents are mutually hostile, whereas the latter seem to be associated with fathers' tendencies to become withdrawn and angry in response to conflict (Katz & Gottman, 1993). Furthermore, mothers' rejection and withdrawal have been related to both types of problems (Fauber, Forehand, Thomas, & Wierson, 1990). Both types of behavior problems involve problems in emotional regulation: lack of regulation of hostile emotions and of aggressive behavior for externalizing behavior and lack of emotional regulation, as evidenced by anxiety and fear, perhaps combined with too high a level of behavioral inhibition, for internalizing behavior (see Eisenberg & Fabes, 1992; Kagan, in press).

The effects of parental conflict and quality of the marital relationship on children appear to be partially mediated by their effects on parenting. For example, low marital satisfaction has been related to negative parenting (i.e., parenting that is cold, unresponsive, and low in limit setting and structure), which in turn predicts children's negative emotion and negative peer interactions (Gottman & Katz, 1989). Moreover, marital conflict has been correlated with inconsistency within and between parents in disciplining daughters

(Stoneman, Brody, & Burke, 1989); in addition, parents preoccupied with their own problems may monitor their children's behavior less than usual. However, as was noted by Grych and Fincham (1990), other mechanisms may underlie the association between parental conflict and child behavior. For example, children may imitate parents' conflict-related actions or become behaviorally unregulated as a consequence of exposure to parental conflict (Cummings & Cummings, 1988; Cummings, Iannotti, & Zahn-Waxler, 1985). As noted previously, Katz and Gottman's (1993) results indicate that marital conflict affects children's regulation of negative emotion. Davies and Cummings (1994) reviewed a substantial body of literature indicating that exposure to adults' conflict, particularly unresolved conflict that is intense or involves a direct threat to the child, often has a dysregulating effect on children's behavior and is associated with children's anger, distress, or behavioral problems. They suggested that parental conflict may affect children's sense of emotional security, and that children's emotional security influences their regulation of emotional arousal and, consequently, their adjustment. Moreover, they argued that conflict may undermine the quality of the parent–child attachment and of parent–child interactions by increasing the negativity of parent–child interactions or decreasing the parent's involvement and accessibility. Moreover, parents engaged in conflict often try to make the child an ally against the other parent (Emery, Joyce, & Fincham, 1987). Decrements in the quality of the parent–child relationship would be expected to undermine children's abilities to regulate their emotion and their adjustment.

Economic Stress. Poverty is a major stressor for many families, particularly for single-parent mother-headed families. Like conflict, economic stress affects familial interactions and children's adjustment; it may even partially account for the negative effects on children of divorce (Amato & Keith, 1991). For example, Conger et al. (1992, 1993) found that economic pressure was associated with maternal and paternal depression, which predicted martial conflict. In addition, the relation between marital conflict and children's adjustment problems was at least partially mediated by the degree to which parenting was uninvolved and nonnurturant or hostile (Conger, Ge, Elder, Lorenz, & Simons, 1994). Other investigators have found that economic hardship affects adolescents' depression/loneliness both directly and indirectly through low nurturance and inconsistent parental discipline; the effects of economic hardship on delinquency and drug use were mediated through inconsistent parental discipline (Lempers, Clark-Lempers, & Simons, 1989). Similar findings were obtained in research on African-American single mothers, although adolescents' perceptions of family economic hardship partially mediated the effects of mothers' financial strain (McLoyd et al., 1994).

Of particular relevance to an understanding of socioemotional development, Brody et al. (1994) found that children's self-regulation mediated the effects of financial resources on rural African-American children's internalizing and externalizing problems. Specifically, the effects of financial resources on rural children's problem behaviors were mediated by parental depression and the quality of the marital relationship, and children's self-regulation mediated the relation between quality of the marital relationship and child problem behaviors. Thus, financial stress appears to affect the quality of the marital relationship, which influences the quality of parenting, and parenting that is nonnurturant or inconsistent is linked to low regulation in children and problems in socioemotional and behavioral functioning.

Summary. Findings such as those just reviewed suggest that changes in the circumstances of families have important consequences for children's socioemotional development. Stress due to intense conflict or economic pressures (which may lead to divorce) often may be associated with inconsistent and harsh parenting practices, diminished quality of the parent–child relationship, and less access for the child to the parent. As a consequence, children's abilities to regulate emotion and behavior are likely reduced, and they are more prone to behavioral problems, including the experience of internalizing and externalizing emotional reactions.

Exposure to adults' negative emotions due to economic stress or marital conflict also may play some role in children's understanding and expression of emotion. If the expression of parental negative emotion in the family is well regulated, resolved, or used as a basis for discussion of feelings, there may be little negative effect on children. However, if, as often is the case, parental emotion is intense, hostile, and/or directed at the child (see Davies & Cummings, 1994), it is likely to be associated with decrements in children's understanding (e.g., Garner et al., 1994) and regulation of negative emotion (e.g., Eisenberg, Fabes, Carlo, Troyer, et al., 1992).

Developmental Considerations

The age at which family stress has the greatest effect on children's socioemotional development is unclear; it likely varies with the nature of the stress, the consequences of the stressor for the structure of the family, the family's affective climate, and parents' practices and meta-emotion philosophy. As argued by Davies and Cummings (1994), children may be vulnerable in different ways at different ages: for example, they may be more vulnerable to aggression, noncompliance, and temper tantrums at an early age whereas dysphoria, passivity, and depression may be more likely outcomes in late childhood and adolescence. Amato and Keith (1991), in a review of the literature on divorce and children's well being, found that decrements in children's functioning associated with divorce seemed to be more marked for children in primary and secondary school, and generally were less for children in the preschool years or college. Elder's (1974, 1979) analysis of the effects of economic deprivation during the Great Depression on children suggests that economic adversity may have greater effects on younger children than older children. Clearly, there is a great need for research on the consequences of familial stress on socioemotional development, the processes that underlie such relations, and how the effects of familial stress vary with developmental level of the child.

INTERVENTIONS

The limited available research suggests that school intervention programs can affect children's social competence, and that part of the effects of such programs may be due to procedures that affect children's understanding and regulation of emotion. For example, the differentiation and identification of positive and negative feelings, as well as the ability to understand how others feel (perspective taking), were among the elements of the interventions used by Spivack, Shure, and colleagues (Spivack & Shure, 1974; Spivack, Platt, & Shure, 1976) to promote young children's socially appropriate behavior. Similarly, Feshbach and Feshbach (1982, 1986) found that an empathy-training program which emphasized cognitive and affective components of empathy resulted in increased prosocial behavior among elementary school children.

Moreover, procedures involving identification and discussion of emotions, as well as emotion management and self-control, are basic to the PATHS curricula being used as an intervention in numerous schools (M. Greenberg, 1996). Children are taught to identify feelings, that all feelings are okay, to label one's own and others' feelings, when to monitor feelings, and how one's behavior affects others' feelings. Moreover, simple procedures for inhibiting behavior when emotionally aroused are taught (Greenberg, Kusche, Cook, & Quamma, 1995). Training has been related to range of children's vocabulary and fluency in discussing emotional experiences, efficacy beliefs regarding the management of emotion, and children's understanding of some aspects of emotion. Moreover, to some degree the intervention was most successful for children high in internalizing or externalizing behavior (Greenberg et al., 1995). In addition, the PATHS curriculum combined with an emphasis on improving relationships between preschool children and their teachers and training problem-solving skills has been related to decreased negative emotion and increased skill with peers and productive involvement in the classroom (Denham & Burton, 1996). Thus, although few researchers have specifically targeted emotional understanding and regulation in interventions aimed at promoting socioemotional competence, initial findings suggest that children's socioemotional competence can be fostered by interventions of this sort.

SUMMARY

Although there probably are constitutional bases to children's dispositional emotionality and regulation, it appears that socializers can promote or undermine children's socioemotional competence in a variety of ways, including through how they react to children's emotions, display emotions, and talk about emotions. Moreover, based on the limited work on the relations of marital and economic stress to children's problem behaviors and skill with peers, it appears that parental depressed emotion and suboptimal parenting behaviors in stressful family situations often may undermine children's socioemotional competence. Thus, an understanding of the socialization of emotion may be particularly important for delineating processes mediating the effects of stress on the family and for designing interventions to enhance children's socioemotional competence.

REFERENCES

Amato, P. M., & Keith, B. (1991). Parental divorce and the well-being of children: A meta-analysis. *Psychological Bulletin, 110,* 26–46.

Balswick, J., & Avertt, C. (1977). Differences in expressiveness: Gender, interpersonal orientation, and perceived parental expressiveness as contributing factors. *Journal of Marriage and the Family, 39,* 121–127.

Batson, C. D. (1991). *The altruism question: Toward a social-psychological answer.* Hillsdale, NJ: Erlbaum.

Block, J. H., & Block, J. (1980). The role of ego-control and ego-resiliency in the organization of behavior. In W. Andrew Collins (Ed.), *Development of cognition, affect, and social relations.* The Minnesota Symposia on Child Psychology (Vol. 13, pp. 39–101). Hillsdale, NJ: Erlbaum.

Boyum, L. A., & Parke, R. D. (1995). Family emotional expressiveness and children's social competence. *Journal of Marriage & Family, 57,* 593–608.

Bridges, L. J., & Grolnick, W. S. (1995). The development of emotional self-regulation in infancy and early childhood. In N. Eisenberg (Ed.), *Review of personality and psychology* (pp. 185–211). Newbury Park: Sage.

Brody, G. H., Stoneman, Z., Flor, D., McCrary, C., Hastings, L., & Conyers, O. (1994). Financial resources, parent psychological functioning, parent co-caregiving, and early adolescent competence in rural two-parent African-American families. *Child Development, 65,* 590–605.

Bronstein, P., Fitzgerald, M., Briones, M., Pieniadz, J., & D'Ari, A. (1993). Family emotional expressiveness as a predictor of early adolescent social and psychological adjustment. *Journal of Early Adolescence, 13,* 448–471.

Bryant, B. K. (1987). Mental health, temperament, family, and friends: Perspectives on children's empathy and social perspective taking. In N. Eisenberg & J. Strayer (Eds.), *Empathy and its development* (pp. 245–270). Cambridge, UK: Cambridge University Press.

Buck, R. (1984). *The communication of emotion.* New York: Guilford Press.

Caspi, A., Henry, B., McGee, R.O., Moffitt, T.E., & Silva, P.A. (1995). Temperamental origins of child and adolescent behavior problems: From age 3 to age 15. *Child Development, 66,* 55–68.

Cassidy, J. (1994). Emotion regulation: Influences of attachment relationships. In N. Fox (Ed.), Emotion regulation: Behavioral and biological considerations. *Monographs of Society for Research in Child Development, 59,* Serial No. 240, 228–249.

Cassidy, J., Parke, R. D., Butkovsky, L., & Braungart, J. M. (1992). Family–peer connections: The roles of emotional expressiveness within the family and children's understanding of emotion. *Child Development, 63,* 603–618.

Coie, J. D., Dodge, K. A., & Kupersmidt, J. B. (1990). Peer group behavior and social status. In S. R. Asher & J. D. Coie (Eds.), *Peer rejection in childhood* (pp. 17–59). Cambridge, UK: Cambridge University Press.

Conger, R. D., Conger, K. J., Elder, G. H., Jr., Lorenz, F. O., Simons, R. L., & Whitbeck, L. B. (1992). A family process model of economic hardship and adjustment of early adolescent boys. *Child Development, 63,* 526–541.

Conger, R. D., Conger, K. J., Elder, G. H., Jr., Lorenz, F. O., Simons, R. L., & Whitbeck, L. B. (1993). Family economic stress and adjustment of early adolescent girls. *Developmental Psychology, 29,* 206–219.

Conger, R. D., Ge, X., Elder, G. H. Jr., Lorenz, F. O., & Simons, R. L. (1994). Economic stress, coercive family process, and developmental problems of adolescents. *Child Development, 65,* 541–561.

Cummings, E. M., & Cummings, J. L. (1988). A process-oriented approach to children's coping with adults' angry behavior. *Developmental Review, 8,* 296–321.

Cummings, E. M., Iannotti, R. J., & Zahn-Waxler, C. (1985). Influence of conflict between adults on the emotions and aggression of young children. *Developmental Psychology, 21,* 495–507.

Davies, P. T., & Cummings, E. M. (1994). Marital conflict and child adjustment: An emotional security hypothesis. *Psychological Bulletin, 116,* 387–411.

Denham, S. A. (1989). Maternal affect and toddlers' social-emotional competence. *American Journal of Orthopsychiatry, 59,* 368–376.

Denham, S.A. (1993). Maternal emotional responsiveness and toddlers' social-emotional competence. *Journal of Child Psychology and Psychiatry, 34,* 715–728.

Denham, S. A., & Burton, R. (1996). A social-emotional intervention for at-risk 4-year-olds. *Journal of School Psychology, 34,* 225–245.

Denham, S. A., Cook, M., & Zoller, D. (1992). 'Baby looks very sad.' Implications of conversations about feelings between mother and preschooler. *British Journal of Developmental Psychology, 10,* 301–315.

Denham, S. A., & Grout, L. (1992). Mothers' emotional expressiveness and coping: Relations with preschoolers' social-emotional competence. *Genetic, Social, and General Psychology Monographs, 118,* 73–101.

Denham, S. A., & Grout, L. (1993). Socialization of emotion: Pathway to preschoolers' emotional and social competence. *Journal of Nonverbal Behavior, 17,* 205–227.

Denham, S. A., Mitchell-Copeland, J., Standberg, K., Auerbach, S., & Blair, K. (in press). Parental contributions to preschoolers' emotional competence: Direct and indirect effects. Motivation and Emotion.

Denham, S. A., Renwick, S. M., & Holt, R. W. (1991). Working and playing together: Prediction of preschool social-emotional competence from mother–child interaction. *Child Development, 62,* 242–249.

Denham, S. A., Renwick-DeBardi, S., & Hewes, S. (1994). Emotional communication between mothers and preschoolers: Relations with emotional competence. *Merrill-Palmer Quarterly, 40,* 488–508.

Denham, S. A., Zoller, D., & Couchoud, E. A. (1994). Socialization of preschoolers' emotion understanding. *Developmental Psychology, 30,* 928–936.

Dunn, J., Bretherton, I., & Munn, P. (1987). Conversations about feeling states between mothers and their young children. *Developmental Psychology, 23,* 132–139.

Dunn, J., & Brown, J. (1994). Affect expression in the family, children's understanding of emotions, and their interactions with others. *Merrill-Palmer Quarterly, 40,* 120–137.

Dunn, J., Brown, J., & Beardsall, L. (1991). Family talk about feeling states and children's later understanding of others' emotions. *Developmental Psychology, 27,* 448–455.

Dunn, J., Brown, J., Slomkowski, C., Tesla, C., & Youngblade, L. (1991). Young children's understanding of other people's feelings and beliefs: Individual differences and their antecedents. *Child Development, 62,* 1352–1366.

Eisenberg, N., & Fabes, R. A. (1990). Empathy: Conceptualization, assessment, and relation to prosocial behavior. *Motivation and Emotion, 14,* 131–149

Eisenberg, N., & Fabes, R. A. (1992). Emotion, regulation, and the development of social competence. In M. S. Clark (Ed.), *Review of personality and social psychology,* Vol. 14. Emotion and social behavior (pp. 119–150). Newbury Park, CA: Sage.

Eisenberg, N., & Fabes, R. A. (1994). Mothers' reactions to children's negative emotions: Relations to children's temperament and anger behavior. *Merrill-Palmer Quarterly, 40,* 138–156.

Eisenberg, N., & Fabes, R. A. (in press). Prosocial development. In W. Damon (Series Ed.) & N. Eisenberg (Vol. Ed), *Handbook of Child Psychology: Vol. 3. Social, emotional, and personality development* (5th ed.). New York: Wiley.

Eisenberg, N., Fabes, R. A., Bernzweig, J., Karbon, M., Poulin, R., & Hanish, L. (1993). The relations of emotionality and regulation to preschoolers' social skills and sociometric status. *Child Development, 64,* 1418–1438.

Eisenberg, N., Fabes, R. A., Carlo, G., Troyer, D., Speer, A. L., Karbon, M., & Switzer, G. (1992). The relations of maternal practices and characteristics to children's vicarious emotional responsiveness. *Child Development, 63,* 583–602.

Eisenberg, N., Fabes, R. A., Carlo, G., & Karbon, M. (1992). Emotional responsivity to others: Behavioral correlates and socialization antecedents. N. Eisenberg & R. A. Fabes (Eds.), *New Directions in Child Development, 55,* 57–73.

Eisenberg, N., Fabes, R. A., & Guthrie, I. (in press). Coping with stress. The roles of regulation and development. In S. Wolchik & I. Sandler (Eds.), *Children's coping: Links between theory and intervention.* New York: Plenum.

Eisenberg, N., Fabes, R. A., Guthrie, I. K., Murphy, B. C., Maszk, P., Holmgren, R., & Suh, K. (1996). The relations of regulation and emotionality to problem behavior in elementary school children. *Development and Psychopathology, 8,* 141–162.

Eisenberg, N., Fabes, R. A., Karbon, M., Murphy, B. C., Wosinski, M., Polazzi, L., Carlo, G., & Juhnke, C. (1996). The relations of children's dispositional prosocial behavior to emotionality, regulation, and social functioning. *Child Development, 67,* 974–992.

Eisenberg, N., Fabes, R. A., Miller, P. A., Shell, C., Shea, R., May-Plumlee, T. (1990). Preschoolers' vicarious emotional responding and their situational and dispositional prosocial behavior. *Merrill-Palmer Quarterly, 36,* 507–529.

Eisenberg, N., Fabes, R. A., & Murphy, B. (1995). The relations of shyness and low sociability to regulation and emotionality. *Journal of Personality and Social Psychology, 68,* 505–517.

Eisenberg, N., Fabes, R. A., & Murphy, B. C. (1996). Parents' reactions to children's negative emotions: Relations to children's social competence and comforting behavior. *Child Development 67*(5), 2227–2247.

Eisenberg, N., Fabes, R. A., Murphy, B., Karbon, M., Maszk, P., Smith, M., O'Boyle, C., & Suh, K. (1994). The relations of emotionality and regulation to dispositional and situational empathy-related responding. *Journal of Personality and Social Psychology, 66,* 776–797.

Eisenberg, N., Fabes, R. A., Murphy, B., Karbon, M., Smith, M., & Maszk, P. (1996). The relations of children's dispositional empathy-related responding to their emotionality, regulation, and social functioning. *Developmental Psychology, 32,* 195–209.

Eisenberg, N., Fabes, R. A., Murphy, M., Maszk, P., Smith, M., & Karbon, M. (1995). The role of emotionality and regulation in children's social functioning: A longitudinal study. *Child Development, 66,* 1239–1261.

Eisenberg, N., Fabes, R. A., Nyman, M., Bernzweig, J., & Pinuelas, A. (1994). The relations of emotionality and regulation to children's anger-related reactions. *Child Development, 65,* 109–128.

Eisenberg, N., Fabes, R. A., Schaller, M., Carlo, G., & Miller, P. A. (1991). The relations of parental characteristics and practices to children's vicarious emotional responding. *Child Development, 62,* 1393–1408.

Eisenberg, N., Fabes, R. A., Schaller, M., Miller, P. A., Carlo, G., Poulin, R., Shea, C., & Shell, R. (1991). Personality and socialization correlates of vicarious emotional responding. *Journal of Personality and Social Psychology, 61,* 459–471.

Eisenberg, N., Guthrie, I. K., Fabes, R. A., Reiser, M., Murphy, B. C., Holmgren, R., Maszk, P., & Losoya, S. (1997). The relations of regulation and emotionality to resiliency and competent social functioning in elementary school children. *Child Development, 68,* 295–311.

Eisenberg, N., & Okun, M. A. (1996). The relations of dispositional regulation and emotionality to elders' empathy-related responding and affect while volunteering. *Journal of Personality, 64,* 157–183.

Eisenberg, N., Schaller, M., Fabes, R. A., Bustamante, D., Mathy, R., Shell, R., & Rhodes, K. (1988). The differentiation of personal distress and sympathy in children and adults. *Developmental Psychology, 24,* 766–775.

Elder, G. H., Jr. (1974). *Children of the depression.* Chicago, IL: University of Chicago Press.

Elder, G. H., Jr. (1979). Economic depression and postwar opportunity in men's lives: A study of life patterns and health. In R. G. Simmons (Ed.), *Research in community and mental health.* Greenwich, CT: JAI Press.

Emery, R. E., Joyce, S. A., & Fincham, F. D. (1987). Assessment of child and marital problems. In K. D. O'Leary (Ed.)., *Assessment of marital discord* (pp. 223–261). Hillsdale, NJ: Erlbaum.

Fabes, R. A., Eisenberg, N., Karbon, M., Bernzweig, J., Speer, A. L., & Carlo, G. (1994). Socialization of children's vicarious emotional responding and prosocial behavior: Relations with mothers' perceptions of children's emotional reactivity. *Developmental Psychology, 30,* 44–55.

Fabes, R. A., Eisenberg, N., & Miller, P. (1990). Maternal correlates of children's vicarious emotional responsiveness. *Developmental Psychology, 26*, 639–648.

Fauber, R., Forehand, R., Thomas, A. M., & Wierson, M. (1990). A mediational model of the impact of marital conflict on adolescent adjustment in intact and divorced families: The role of disrupted parenting. *Child Development, 61*, 1112–1123.

Feshbach, N. D., & Feshbach, S. (1982). Empathy training and the regulation of aggression: Potentialities and limitations. *Academic Psychology Bulletin, 4*, 399–413.

Feshbach, S., & Feshbach, N. D. (1986). Aggression and altruism: a personality perspective. In C. Zahn-Waxler, E. M. Cummings, & R. Iannotti (Eds.), *Altruism and aggression: Biological and social origins* (pp. 189–217). Cambridge, UK: Cambridge University Press.

Garner, P. W., Jones, D. C., & Miner, J. L. (1994). Social competence among low-income preschoolers: Emotion socialization practices and social cognitive correlates. *Child Development, 65*, 622–637.

Gianino, A., & Tronick, E. Z. (1988). The mutual regulation model: The infant's self and interactive regulation, coping, and defense. In T. Field, P. McCabe, & N. Schneiderman (Ed.), *Stress and coping* (pp. 47–68). Hillsdale, NJ: Erlbaum.

Gottman, J. M., & Katz, L. F. (1989). Effects of marital discord on young children's peer interaction and health. *Developmental Psychology, 25*, 373–381.

Gottman, J. M., Katz, L. F., & Hooven, C. (1996). Meta-emotion and socialization of emotion in the family — A topic whose time has come: Comment on Gottman et al. (1996). *Family Psychology, 10*, 269–276.

Greenberg, M. T. (1996). The PATHS project: Preventive intervention for children. University of Washington; Final report to NIMH.

Greenberg, M. T., Kusche, C. A., Cook, E. T., & Quamma, J. P. (1995). Promoting emotional competence in school-aged children: The effects of the PATHS curriculum. *Development and Psychopathology, 7*, 117–136.

Grusec, J. E., & Goodnow, J. J. (1994). Impact of parental discipline methods on the child's internalization of values: A reconceptualization of current points of view. *Developmental Psychology, 30*, 4–19.

Grych, J. H., & Fincham, F. D. (1990). Marital conflict and children's adjustment: A cognitive–contextual framework. *Psychological Bulletin, 108*, 267–290.

Halberstadt, A. G. (1986). Family socialization of emotional expression and nonverbal communication styles and skills. *Journal of Personality and Social Psychology, 51*, 827–836.

Hardy, D. F., Power, T. G., & Jaedicke, S. (1993). Examining the relation of parenting to children's coping with everyday stress. *Child Development, 64*, 1829–1841.

Hoffman, M. L. (1983). Affective and cognitive processes in moral internalization. In E. T. Higgins, D. N. Ruble, & W. W. Hartup (Eds.), *Social cognition and social development: A sociocultural perspective* (pp. 236–274). Cambridge, MA: Cambridge University Press.

Kagan, J. (in press). Biology and the child. In W. Damon (Series Ed.) & N. Eisenberg (Vol. Ed), *Handbook of Child Psychology: Vol. 3. Social, emotional, and personality development* (5th ed.). New York: Wiley.

Katz, L. F., & Gottman, J. M. (1993). Patterns of marital conflict predict children's internalizing and externalizing behavior. *Developmental Psychology, 29*, 940–950.

Kestenbaum, R., Farber, E. A., & Sroufe, L. A. (1989). Individual differences in empathy among preschoolers: Relation to attachment history. In N. Eisenberg (Ed.), *New directions for child development: Vol. 44. Empathy and related emotional responses* (pp. 51–64). San Francisco: Jossey-Bass.

Laird, R. D., Pettit, G. S., Mize, J., Brown, E. G., & Lindsey, E. (1994). Mother–child conversations about peers: Contributions to competence. *Family Relations, 43*, 425–432.

Larsen, R. J., & Diener, E. (1987). Affect intensity as an individual difference characteristic: A review. *Journal of Research in Personality, 21*, 1–39.

Lempers, J. D., Clark-Lempers, D., & Simons, R. L. (1989). Economic hardship, parenting, and distress in adolescence. *Child Development, 60*, 25–39.

Maccoby, E. E., & Martin, J. A. (1983). Socialization in the context of the family: Parent–child interaction. In P. H. Mussen (Ed.), *Handbook of child psychology. Vol IV. Socialization, personality, and social development* (E. M. Hetherington, Ed.; pp. 1–101). NY: Wiley.

Malatesta, C. Z., & Haviland, J. M. (1985). Signals, symbols, and socialization: The modification of emotional expression in human development. In M. Lewis & C. Saarni (Eds.), *The socialization of emotions* (pp. 89–116). NY: Plenum.

Matthews, K. A., Woodall, K. L., Kenyon, K., & Jacob, T. (1996). Negative family environment as a predictor of boys' future status on measures of hostile attitudes, interview behavior, and anger expression. *Health Psychology, 15*, 30–37.

McLoyd, V. C., Jayaratne, T. E., Ceballo, R., & Borquez, J. (1994). Unemployment and work interruption among African American single mothers: Effects on parenting and adolescent socioemotional functioning. *Child Development, 65*, 562–589.

Olweus, D. (1980). Familial and temperamental determinants of aggressive behavior in adolescent boys: A causal analysis. *Developmental Psychology, 16*, 644–666.

Plomin, R., & Stocker, C. (1989). Behavioral genetics and emotionality. In J. S. Reznick (Ed.), *Perspectives on behavioral inhibition* (pp. 219–240). Chicago: Chicago University Press.

Pulkkinen, L. (1982). Self-control and continuity from childhood to late adolescence. In P. B. Baltes & O. G. Brim, Jr. (Eds.), *Life-span development and behavior* (Vol. 4.) New York: Academic.

Roberts, W., & Strayer, J. (1987). Parents' responses to the emotional distress of their children: Relations with children's competence. *Developmental Psychology, 23*, 415–432.

Robins, R. W., John, O. P., Caspi, A., Moffitt, T. E., & Stouthamer-Loeber, M. (1996). Resilient, overcontrolled, and undercontrolled boys: Three replicable personality types. *Journal of Personality and Social Psychology, 70*, 157–171.

Rothbart, M. K., & Bates, J. E. (in press). Temperament. In W. Damon (Series Ed.) & N. Eisenberg (Vol. Ed), *Handbook of Child Psychology: Vol. 3. Social, emotional, and personality development* (5th ed.). New York: Wiley.

Rubin, K. H., Bukowski, W., & Parker, J. G. (in press). Peer interactions, relationships, and groups. In W. Damon (Series Ed.) & N. Eisenberg (Vol. Ed), *Handbook of child psychology: Vol. 3. Social, emotional, and personality development* (5th ed.). New York: Wiley.

Shields, S. A. (1995). The role of emotion beliefs and values in gender development. In N. Eisenberg (Ed.), *Review of personality and social psychology. Vol. 15. Social development* (pp. 212–232). Thousand Oaks, CA: Sage.

Spivack, G., & Shure, M. B. (1974). *Social adjustment of young children*. San Francisco: Jossey-Bass.

Spivack, G., Platt, J. J., & Shure, M. B. (1976). *The problem solving approach to adjustment*. San Francisco, CA: Jossey-Bass.

Sroufe, L. A., Schork, E., Motti, F., Lawroski, N., & LaFreniere, P. (1984). The role of affect in social competence. In C. E. Izard, J. Kagan, & R. B. Zajonc (Eds.), *Emotion, cognition, and behavior* (pp. 289–319). Cambridge, UK: Cambridge University Press.

Stoneman, Z. Brody, G. H., & Burke, M. (1989). Marital quality, depression, and inconsistent parenting: Relationship with observed mother–child conflict. *American Journal of Orthopsychiatry, 59*, 105–117.

Strauss, C. C. (1988). Social deficits of children with internalizing disorders. In B. B. Lahey & A. E. Kazdin (Eds.), *Advances in clinical child psychology. Vol. 11* (pp. 159–191). NY: Plenum.

Thompson, R. (in press). Early sociopersonality development. In W. Damon (Series Ed.) & N. Eisenberg (Vol. Ed), *Handbook of child psychology: Vol. 3. Social, emotional, and personality development* (5th ed.). New York: Wiley.

Tomkins, S. S. (1963). *Affect, imagery, consciousness. Vol. 2 Negative affects*. New York: Springer.

Zeman, J., & Shipman, K. (1996). Expression of negative affect: Reasons and methods. *Developmental Psychology, 32*, 842–849.

5

THE SOCIALIZATION OF CHILDREN'S EMOTIONAL AND SOCIAL BEHAVIOR BY DAY CARE EDUCATORS[*]

Donna R. White and Nina Howe

Concordia University

One of the most dramatic changes in the lives of children in the last three decades has been the increasingly widespread use of group day care in the preschool and school-age years. Preschool care was originally viewed as an intervention for improving academic readiness of children from low socio-economic families. Early research comparing home-reared and group care children from economically disadvantaged or middle class families indicated that preschool child care facilitated intellectual development in disadvantaged children (e.g., Ramey, Bryant, Campbell et al., 1990) and was not detrimental to development in children from more advantaged families (e.g., Belsky & Steinberg, 1978; Doyle, 1975; Etaugh, 1980).

Given the large increment in the number of women in the North American work force, more infants, toddlers, preschool and school-age children from all economic levels are being cared for in group settings (Hofferth, 1992; Kamerman & Kahn, 1981). Current research on child care has also changed in focus. First, the goal of child care is no longer seen solely as academic preparation. It is recognized that emotional and social development are important in the early years, and that parents want their children to be in warm, secure environments where emotional regulation and social competence with peers and adults are fostered. Secondly, it has been recognized that both day care and home care are extremely variable in quality and each could be seen as a positive or negative influence on child development. The type of care provided in both group and home settings can facilitate or retard child development. Furthermore, day care versus home care comparisons have not provided answers to questions involving how to improve children's experience in group settings. The notion of quality, assessed by global measures (Harms & Clifford,

* Writing of this chapter was supported by grants from the Child Care Initiatives Fund, Health and Welfare Canada (Project no: 4774-5-91/9-688) and Child Care Visions, Canada Human Development (Project no: 4561-15-96/074) to the first author and the Social Sciences and Humanities Research Council of Canada, Concordia University Research Fund, and Fonds pour la Formation de Chercheurs et l'Aide à la Recherche (Québec) to the second author.

Improving Competence across the Lifespan, edited by Pushkar *et al.*
Plenum Press, New York, 1998.

1980; Harms, Jacobs, & White, 1995) or regulatable characteristics (e.g., low teacher child ratios, smaller group size) helped to clarify the idea that high quality care was related to positive development outcomes (White & Mill, in press). However, once again, such studies did not provide sufficient direction for understanding the processes leading to optimal child development in day care contexts. More recently, researchers, educators, and parents have agreed that to understand children's experience in child care, it is necessary to focus on educator–child and peer interactions (Kontos & Wilcox-Herzog, 1997).

In this chapter, we will briefly review the results of studies investigating home care versus day care and quality of care comparisons on the emotional and social development in preschool and young school-age children. We will then turn to other sources in order to discuss how educators might facilitate emotional and social development in children. First, as demonstrated in the accompanying chapter by Nancy Eisenberg, there is an extensive body of literature on parent socialization of children's socioemotional competence. The results of work reviewed by Eisenberg provide a model for teacher–child interactions and their influence on socioemotional development. Second, work on maternal socialization style and preschool-aged siblings' caretaking of toddler-aged siblings (Howe & Rinaldi, 1996; 1997) will be reviewed and related to day care educators' socialization of behaviour toward peers.

We will next turn to studies of educator behaviour toward children in day care. It seems likely that educators will employ similar socialization strategies as mothers and that much of the research on maternal socialization can be directly applied to educator behaviour. Nonetheless, children who attend group day care are exposed to an environment that differs from home care in a number of ways. There is little research directed towards understanding how differences such as exposure to multiple caregivers and to larger groups of peers might effect emotional regulation and social development. In this context, we will examine the effects of child care by reviewing work on four aspects of educator behaviour: caregiver attachment and stability, group size/educator–child ratio, caregiver warmth, and educator discussion of feelings and interactions. Finally, we will try to apply the parent and sibling work and the work on educator behaviour in the group environment to the formulation of guidelines for educators and directions for future research.

DAY CARE VERSUS HOME CARE COMPARISONS OF SOCIAL AND EMOTIONAL DEVELOPMENT

Early studies of the influence of day care on socioemotional development produced mixed results. Although findings often pointed to a general pattern of enhanced social competence and peer interactions (see reviews by Belsky & Steinberg, 1978; Clarke-Stewart & Fein, 1983; Etaugh, 1980), center-based children were also observed to be more aggressive with peers and less compliant with adults than their home-reared counterparts (Haskins, 1985).

Few studies have addressed the relationship of school-age care and socioemotional development in children. Once again, children from 5 to 12 years are exposed to greater peer contact and multiple caregivers when they attend school-age care programs. In a longitudinal study, White (1995) found that attendance in school-age care programs was related to greater aggression in kindergarten-age children, even after the effects of preschool group care were controlled. Adessky (1996) noted that hours of group participation in kindergarten, after-school French programs, and school-age care programs were related to increased aggression in middle class boys, but not girls.

It may well be that group environments are related to both social competence and aggressive behaviour. Children who spend time in a group setting must be able to negotiate with peers and win friends. At times, they may also need to "stick up for themselves" in terms of obtaining materials or not letting others victimize them. On the other hand, these early studies did not take quality of the child care environment into account and the mixed results may reflect differences in quality of care from study to study, with high quality care facilitating positive social behaviors and low quality care increasing aggression.

STUDIES OF GLOBAL QUALITY AND SOCIOEMOTIONAL DEVELOPMENT

In order to determine if the inconsistent findings of home care versus day care studies were a function of methodological problems related to heterogeneity of care within each group, studies of global quality of the child care environment were conducted. A commonly used measure of global quality at the preschool level is the Early Childhood Environment Rating Scale (Harms & Clifford, 1980), in which observers rate the materials, use of space, activities, and other aspects of the day care environment. Preschool children in high quality centers were found to exhibit greater social competence than children in low quality centers in several studies. Specifically, children in high quality centers show greater consideration and sociability (Phillips, McCartney, & Scarr, 1987; Phillips, Scarr, & McCartney, 1987), engage in more cooperative behaviour (Howes & Olenick, 1986), and in less negative play and fewer negative peer interactions (White, Jacobs, & Schliecker, 1988) compared to children in lower quality settings.

Studies of quality of school-age centers and socioemotional development are not yet available. In 1995, a global quality scale for school-age care (Harms, Jacobs, & White, 1995) was published and should provide the basis for investigating the relations between school-age environment quality, emotional regulation, and social competence.

While the preschool studies support the notion that "high global quality" is related to better social and emotional development, they do not provide much guidance regarding how educators might facilitate social competence and emotional regulation. It is difficult to find research on child care educators as socializing agents, but, as illustrated in the accompanying chapter by Nancy Eisenberg, there is a large body of literature on maternal socialization of children's socioemotional competence.

In the sections that follow, we highlight some of Eisenberg's main points, and discuss recent work by Howe and Rinaldi (1996, 1997), who investigated maternal socialization of socioemotional responses toward distressed siblings. It must be noted that although references are made to adults and parents as socializing agents, nearly all of the research has focused on mother–child interactions. Nonetheless, it seems reasonable that what we have learned from studies of maternal socialization may be applied to day care educators, and that maternal-sibling studies may lead to some ideas about how educators teach children to manage peer interactions.

MATERNAL SOCIALIZATION OF SOCIOEMOTIONAL COMPETENCE

We are all emotional beings, although the degree to which we express our emotions and our ability to regulate emotions in appropriate ways in different social situations varies widely across individuals. Certainly some of these individual differences are due to temperamental dispositions, however, it is clear that parents play an important role in promoting young children's socioemotional development. In this volume, Eisenberg argues that adults may facilitate or undermine children's socioemotional competence in a number of ways, and in particular, identified three critical factors: (a) how mothers react to children's emotional displays, (b) how mothers themselves exhibit and cope with emotions, and (c) how mothers talk about emotions to children, and thus teach children to regulate their emotions. Parental socialization typically occurs within the context of on-going interaction with others. The interactional and relational context of socialization is important because it allows children to develop an understanding of particular emotions and how specific contexts (e.g., separation) may elicit particular emotional responses, which in turn, they need to learn to regulate or control effectively.

The responses of socializing agents such as parents to children's emotions appear to affect the degree to which children both express and regulate their own emotions. By regulating emotions we refer to the ability of children to experience and express positive emotions such as happiness, as well as control negative feelings such as anger or frustration; specifically the processes of initiating, maintaining, and moderating emotional responsiveness (Grolnick, Bridges, & Connell, 1996). Negative adult reactions to children's emotions have been linked to the greater likelihood that children will hide their own emotions and feel anxious, whereas positive adult reactions to emotions have been associated with children's ability to express emotions without shame and discomfort. Although, as Eisenberg points out, there are situations in which it may be socially appropriate to learn to restrict or inhibit one's true feelings, for example, learning not to stare at a handicapped person. However, researchers have generally reported that maternal encouragement of children's expression of emotion is associated with positive outcomes, such as a greater understanding of emotions and better peer relations.

One of the major lessons that children learn via maternal reactions to emotions are specific means to deal with and control their own emotions. Through modelling and encouragement by parents, children can learn to inhibit and control some emotional displays appropriately (particularly negative ones), seek social support, and acquire skills for learning to cope in instrumental ways. An example of coping in an instrumental way would be talking about an emotional situation such as why the child is frustrated by completing a difficult puzzle and then generating solutions to the problem. In contrast, crying in frustration, throwing puzzle pieces, and giving up would be noninstrumental or immature responses.

Adults' expression of emotions certainly influences children's socioemotional competence and in fact, there are similarities in the degree to which parents and their children express both positive and negative emotions. For example, the more that parents are able to express empathy, the more likely their children will also be empathetic (e.g., Eisenberg, Fabes, Schaller, Carlo, & Miller, 1991).

Another critical task is learning how to understand emotions both in the self and others. For example, learning to regulate negative emotions successfully in family situations has been linked positively to socioemotional competence, but primarily when it is accompanied by adult–child discussion of emotions. In the process of talking about emotional expressions and situations that elicit emotional responses, children can learn to identify and label specific feelings and understand the nature of emotional knowledge (e.g., causes and functions of emotions). A number of authors such as Dunn (1988) and Harris (1994) argue that family membership facilitates social knowledge, that is children develop an understanding of other's feelings, thoughts, and intentions within the intimate and special context of on-going family interactions. Moreover, the kind of social knowledge and emotional understanding developed within the family context presumably will transfer to children's interactions with peers and other adults in settings outside of the home. In fact, family discourse about emotions has been positively related to children's affective understanding of others, as well as more positive peer and sibling relations (see Dunn, 1993). Recently, Brown, Donelan-McCall, and Dunn (1996) reported that children who employed emotional language during play had more cooperative peer and sibling relations and performed at a more sophisticated level on a cognitive task measuring children's beliefs of what others thought.

In conclusion, there is evidence to support the notion that discussion of emotional states is an important process implicated in the development of positive social relations. Mothers may also teach appropriate peer behaviour in the context of the sibling relationship. We present some data from a study on maternal socialization of emotions and sibling caretaking that will serve as a springboard for a discussion of how parental and sibling models might extend our understanding of the educator's role in children's socioemotional socialization.

SIBLING CARETAKING AND EMOTIONAL SOCIALIZATION

In recent work (Howe & Rinaldi, 1996, 1997), the role of maternal socialization of emotional behaviour and the ability of preschool-aged siblings to engage in caretaking of their toddler-aged siblings was investigated. In particular, the investigators were interested in whether hypothetical or real situations were more strongly related to children's practical emotional knowledge about caretaking.

Thirty-two mothers and sibling pairs (preschool and toddler-aged) participated. First, the emotional knowledge of mothers and older siblings was measured by the Parent–Child Affect Communication Task or PACT (adapted from Zahn-Waxler, Ridgeway, Denham, Usher, & Cole, 1993); mothers and children looked at eight pictures of infants in different emotional states (e.g., happy, sad, distressed) and their conversations about the pictures were tape-recorded and transcribed. Their conversations were coded for references to internal states (emotions, cognitive states, intentions), physical manifestations of emotions (e.g., smiling), and causes and functions of emotions. The PACT was the hypothetical measure of maternal emotional socialization. Second, families came to a university playroom where the ability of the older sibling to engage in caretaking of the toddler was assessed. To measure caretaking, the siblings were exposed to a situation that was expected to be moderately stressful for some pairs. Specifically, after the families had been in the playroom for five minutes, mothers were asked to leave the siblings alone for 2.5 minutes (mothers continued to observe their children from behind a one-way mirror). The mother–older sibling conversations at the moment of leave-taking were coded for references to the toddler's internal states and were the measure of real-life maternal emotional socialization. After mother departed, if the younger child became upset (e.g., cried, whined, searched), the older sibling's caretaking behaviour was coded (e.g., patting, kissing, soothing statements, giving toys).

One question examined was the impact of maternal emotional socialization during hypothetical (discussion of the PACT) and real situations (discussion at leave taking) on sibling caretaking. Mothers demonstrated a consistency of style in terms of emotional socialization, that is, mothers who talked about internal states on the PACT also did so at leave-taking in the stressful situation ($r = .40$, $p < .05$). However, contrary to expectations, both maternal and older sibling discussion on the PACT were negatively related to sibling caretaking ($r = -.34, -.32, p < .05$, respectively). That is, when mothers and children talked about the emotional expressions depicted in the eight pictures of enfants, older siblings were less likely to comfort their younger siblings in the laboratory situation. This finding is in line with Garner, Jones, and Miner's (1994) report of a negative association between negative emotional expressiveness with the family and sibling caretaking. On the PACT measure, 5 of 8 infant pictures were of negative emotions, thus it may be biased in this direction. Perhaps the older siblings had not yet internalized a link between identifying specific internal states, particularly negative ones, and their causes and the ability to act in a potentially stressful situation. This interpretation is consistent with Howe and Rinaldi's finding that there was a significant, positive relationship between maternal socialization in the real leave-taking situation and sibling caretaking. Specifically, mothers who made references to the toddler's internal states during leave-taking in the laboratory caretaking session had older children who engaged in caretaking or comforting their distressed young sibling ($r = .33, p < .05$). In this emotionally-arousing situation, maternal discussion of the toddler's state may have provided important clues that facilitated the older sibling's understanding of the toddler's emotional state (i.e., that the toddler might cry). Mothers also appeared to give information about emotional regulation to their older children and made comments such as:

> **Family 10**: *"You take care of Linda* (younger sibling), *OK? Make sure she doesn't start crying?"*
> **Family 21**: *"You keep Jenny* (younger sibling) *busy, OK?"*

Some mothers probably anticipated that their toddlers would become distressed and part of their instructions were related to distracting this child by keeping both children actively engaged with the toys in the playroom. Others have reported that similar strategies, such as keeping busy, were positive ways for young children to regulate negative emotions in mildly stressful situations (Grolnick et al., 1996).

In fact, adults may need to teach children what to do in particular types of emotional situations through a process called scaffolding. That is, by getting the child's attention (i.e., the toddler might cry) and providing an appropriate kind of response for the older sibling (e.g., keep the toddler busy, play with or comfort him/her), the mother may set the stage for a prosocial response by an older sibling. Denham, Mason, and Couchoud (1995) reported that preschoolers were more likely to respond prosocially to adult's emotions after supportive scaffolding by the adult. Two aspects of scaffolding, labelling/explaining the adult's emotion (sadness, anger, or pain) and requesting prosocial intervention (i.e., asking for help), both facilitated preschoolers' prosocial responses. Younger preschoolers benefitted from the adult scaffolding more than older ones suggesting that as children develop, their emotional understanding becomes more sophisticated. In conclusion, by talking about emotions, adults may play a critical role in fostering children's emotional understanding and optimal socioemotional development. Moreover, mothers apparently can directly encourage prosocial behaviour through suggestion and scaffolding techniques, a finding that is important for our understanding of how day care educators might engage in emotional socialization.

WHAT CAN EDUCATORS LEARN FROM RESEARCH ON MATERNAL SOCIALIZATION?

Proposed Similarities in Maternal and Educator Socialization. Given the amount of time that many children now spend in child care and after-school care settings, it is important to ask if educators guide social interactions and emotional regulation using the same strategies as mothers. We consider whether the model of emotional socialization, as outlined in the accompanying chapter by Nancy Eisenberg and work by Howe and Rinaldi (1996; 1997) described in this chapter, can provide a basis for understanding the dynamics of children's lives in child care. In the day care setting, we can consider the behaviour of the educator from the three major factors outlined by Eisenberg and consider how they might influence the behaviour and emotional knowledge and understanding of the children in their care.

We would expect that a similar relationship between these three dimensions of behaviour would exist between educators and social competence and emotional regulation in children. Thus, educators who encourage children's emotional expression, who model the expression of positive emotions, and who talk about emotions with children would be expected to foster the ability of the children in their care to engage in emotional regulation and socially competent behaviors with educators and peers. None of these dimensions have been examined in much detail in child care contexts, nevertheless the daily life of a child in day care is filled with emotional and stressful situations (e.g., separation from parents, sharing with peers, waiting your turn to be served at the lunch table). Since young children who are able to express both positive and negative emotions appropriately are better liked by their peers (Sroufe, Schork, Motti, Lawroski, & LaFreniere, 1984), we would expect that a major task of the educator would be to teach children how to regulate and express their emotions appropriately within the context of on-going peer interactions and other situations that arise in day care. As indicated by Howe and Rinaldi (1996; 1997), mothers can foster socially competent behaviour towards siblings. Thus, we would expect that educators can also teach children to deal more effectively with peers and stressful situations by using scaffolding techniques.

Certainly, we would expect that when young children spend long hours in child care settings or in after-school programs, the kinds of socialization practices that educators employ would have an impact on children's socioemotional development. As a case in point, separation is a major issue for parents, children, and educators in child care. Parental concerns and anxieties regarding placing their child in the care of a non-family member may be high, even when parents generally feel comfortable with the decision to use group child care. Many children experience distress at being left in the care of others, although typically the distress is short-lived. As any educator can report, children usually stop crying within a few minutes of the departure of their parents and settle into a happier and more emotionally balanced state and begin to interact with peers. Unfortunately, the last image with which parents may leave the day care is that of an unhappy, miserable, angry child. This image may facilitate an increase of guilt and anxiety on the part of the parent. The question of how both parents and educators deal with the actual moment of separation does not seem to have been addressed in the research literature. Specifically, how the educator reacts to the child's emotional strain of parental separation has not been examined, nor the emotional response of educators to separation, or how they talk about the stressful situation to the child and parent. Finally, the issue of how educators facilitate children's emotional regulation following parental separation has not been investigated. Following from the Howe and Rinaldi work, we speculate that children's distress would be moderated by both parents' and educators' sensitivity to the child's emotional state. This sensitivity would be manifested in a discussion of how the child is feeling, why he/she is in the specific emotional state, and the educator's attempts to comfort and distract the child. Presumably, a warm, calm, empathic reaction on the part of the educator to both distressed parents and children would facilitate their ability to regulate their negative emotions and to regain a positive emotional state (e.g., a busy, happy child and a parent who goes to work without guilt and anxiety).

Unique Features of the Child Care Environment and Their Influence on Educator–Child Interactions. There are a number of critical differences between parents and child care educators that must be factored into any discussion of the feasibility of applying a parental model of socialization to how teachers facilitate children's socioemotional development. For most children, their parents are usually the stable, consistent, and primary caretakers in their lives and thus, we assume, play the critical role in their socialization. Although many children spend long hours in the care of other individuals (ranging from relatives, neighbours, to professionally trained educators), parents still retain the primary responsibility for their upbringing, a fact recognized by both educators and children. For example, although children often develop stable and close, secondary attachments to their caretakers, their primary attachment is still to their parents (e.g., Ainslie & Anderson, 1984; Howes, Rodning, Galluzzo, & Myers, 1988). Thus, the attachment to the educator is not replaced by that to their parents, although there is evidence that a secure, secondary attachment to an educator may partially compensate for a poor mother–child attachment relationship (Howes et al., 1988). Given the primary attachment relationship with parents, it is probable that they may have greater influence on socioemotional development than educators. On the other hand, the number of hours spent in child care and the availability of peers may provide more and different socialization contexts for the educator, giving weight to the educator's socialization practices.

There are also critical differences in adult/child ratios that must be highlighted. Children raised in the typical two-child family are likely to receive more frequent attention from their parent(s) than children in child care where the adult/child ratio may be 1:8 or more in a classroom. Moreover, the actual size of the group will influence the frequency and type of adult–child interactions in child care, with fewer interactions in larger groups (Kontos & Wilcox-Herzog, 1997). In addition, in group child care children typically have multiple educators, as a result of staffing schedules, more than one teacher to a group, yearly changes in classrooms, and staff turn-over.

Nevertheless we argue that child care settings provide a natural and safe environment for children to develop into highly competent socioemotional individuals. This may be true for a number of reasons. Clearly, the dynamics of living in group situations with a large number of peers requires a number of sophisticated social skills such as taking turns, sharing, negotiating play scenarios and conflict resolutions, exchanging information in a respectful manner, leadership, compromising, recognizing others' rights and points of view, and engaging in humorous and playful exchanges. Of course, all of these interactions require children to demonstrate and act in appropriate ways if their social interactions are to be considered prosocial and not aggressive or aversive to others. As we know, one of the major tasks of childhood is to learn to express and regulate both positive and negative emotions in socially acceptable ways. Certainly, child care educators spend considerable time and energy managing the emotional climate of their classroom and by implication, the socioemotional interactions of the children in their care.

THE ROLE OF THE EDUCATOR IN FACILITATING SOCIOEMOTIONAL DEVELOPMENT

Based upon the preceding discussion, it seems likely that day care educators play an important role in socioemotional development through several mechanisms. First, security of caregiver attachment and caregiver stability would seem to be precursors of emotional regulation. Secondly, educators work in a unique environment, caring for several children in groups that are larger than most families. In this context, educators must create a warm, caring environment in order to allow children to development emotional regulation skills and social competence. Finally, educators may influence socioemotional development by the ways they react to, model both emotions and coping responses, discuss emotions with children, and through directly teaching appropriate social behaviour by means of scaffolding and monitoring of on-going interactions. In the following sections, we examine available research data linking educator behaviour to children's socioemotional development.

Caregiver Stability and Attachment. Attachment theory predicts that stable caregivers are essential in forming positive emotional bonds, the basis of socioemotional development, with children. Yet, day care studies have consistently found high rates of staff turnover. Estimates of staff change range from 20% to 70% in a single year in American centers (Hartman & Pearce, 1989; Phillips, Howes, & Whitebook, 1991). According to Phillips et al. (1991), poor wages are the strongest predictor of turnover of child care providers. Poor support offered to staff, poor co-worker relations, lack of preparation time, lack of opportunities for advancement, and lack of space for meeting adult needs are also related to high turnover rates. Given that a relationship between high rates of staff turnover and poor child development has been reported (Kagan, Kearsley, & Zelazo, 1976) and that children are able to develop positive peer relations and social competence in contexts in which they have secure attachment to caregivers (Hayes et al., 1990), the issue of turnover is an important one.

Group Size and Teacher–Child Ratio. Results of a study by Howes, Smith, and Galinsky (1995 cited by Kontos & Wilcox-Herzog, 1997) revealed that lower educator–child ratios were related to warmer, more responsive educator behaviour. In addition, moderate group size has been found to be related to positive peer relations and social competence (Hayes et al., 1990). Such data support the need to keep group size moderate and educator–child ratios low in order to allow for more positive interactions and to help educators facilitate socioemotional development.

Educator Warmth and Harshness. We have argued that socioemotional development will be fostered when educators engage in warm, caring interactions rather than harsh, critical reactions. Is there any data to support the idea that educators display significant amounts of warmth to children in their care or that educator warmth is related to social and emotional development? Data relating educator warmth and socioemotional development is sparse, but Kontos and Wilcox-Herzog (1997) cite several studies indicating that sensitive (warm) teachers respond quickly to children in distress (e.g., crying) compared to detached teachers who are not responsive or harsh teachers who are more likely to be critical and punish the child. Mill (1997) also observed educator–child interactions and found that on average educators displayed 124 affectionate acts (e.g., smiling, affectionate words) toward children per hour. In contrast, only 12% of the educators displayed angry or harsh behaviour towards children. There is some evidence that when children had more frequent warm interactions with their teachers, they were more likely to exhibit strong positive affect (Hestenes, Kontos, & Bryan, 1993). As well, attentive and encouraging teachers facilitated reduced stress in children compared to punitive, critical and emotionally detached teachers (Love, Ryer, & Faddis, 1992). Clearly, the educator's degree of warmth and sensitivity can play an important role in facilitating children's socioemotional development.

Educator Behaviour and Discussion of Feelings and Interactions. Very few studies have investigated the educators' reactions to emotions expressed by children, educator modelling of emotional behaviour, educator–child discussion of emotions, or educator scaffolding in social interactions. Kontos and Herzog-Wilson (1997) briefly described two studies that appear to support the idea that discussion about feelings is rare in child care settings. For example, McCartney (1984) found that caregivers generally did not talk to children about emotions, while Kontos and Dunn (1993) found that educator talk focused on praise, nurturance, redirection or limit setting rather than emotional regulation or socially appropriate behaviour. Clearly, based upon the work with mothers reviewed in this chapter, more research is needed on educator discussion of children's feelings and interactions, as well as on educator reactions, modelling of emotions, and scaffolding.

APPLICATIONS TO THE CLASSROOM AND DIRECTIONS FOR FUTURE RESEARCH

In order to foster socioemotional development, it would seem important to promote educator–child attachment in day care settings by reducing staff turnover and keeping educator–child ratios and group size reasonably low. Studies indicate that educators' chief reasons for leaving their jobs are low wages, low prestige accorded to the profession, and poor working conditions (Phillips et al., 1991). It would seem essential that government policy support higher wages and better working conditions for child care educators.

Another factor promoting socioemotional development is the educator's ability to create a warm environment and to interact with the children in ways that are positive rather than harsh or negative. Interestingly, Mill (1997) found that educator warmth was linked to other quality indicators. Educators who were in high quality environments tended to react more warmly to children. Mill (1997) also found that when educators were harsh with children in their care, they perceived supervisor support to be poor. Policy makers should note that the quality of the environment may facilitate the ability of educators to interact warmly with children, and that increased support to these caregivers may help to reduce angry behaviors they direct towards children.

It is important that educators acknowledge and come to understand that their reactions, their own behaviour and what they say to children about emotions and interactions are important. This area has received virtually no attention by researchers. Research is needed to determine if children's socioemotional development is facilitated by educators in ways that are similar to parents' techniques. The unique conditions of the child care setting must also be studied in relation to children's development of emotional regulation and social competence. It is essential that research provide direction to educators and help them to know what to say and what to do with children in these situations. Such work should be done with both preschool and school-age children who spend much of their time in group care. We can raise salaries and offer more support, but in the final analysis, it is the educator's behaviour and interactions with children that are critical in social and emotional development of children in their care.

REFERENCES

Adessky, R. (1996). The relationship of group and family experiences to peer rated aggression and popularity in middle class kindergarten children. Doctoral Thesis, Department of Psychology, Concordia University, Montreal, Quebec.

Ainslie, R., & Anderson, C. (1984). Daycare children's relationships to their mothers and caregivers: An inquiry into the conditions for the development of attachment. In R. Ainslie (Ed.), *The child and the daycare setting: Qualitative variations and development* (pp. 98–131). New York: Praeger.

Belsky, J., & Steinberg, L.D. (1978). The effects of day care: A critical review. *Child Development, 49*, 929–949.

Brown, J. R., Donelan-McCall, N., & Dunn, J. (1996). Why talk about mental states? The significance of children's conversations with friends, siblings, and mothers. *Child Development, 67*, 836–849.

Clarke-Stewart, K. A., & Fein, G. G. (1983). Early childhood programs. In P.H. Mussen (Ed.), *Handbook of child psychology* (Vol. 2 , pp. 917–999). New York: Wiley.

Denham, S. A., Mason, T., & Couchoud, E. A. (1995). Scaffolding young children's prosocial responsiveness: Preschoolers' responses to adult sadness, anger, and pain. *International Journal of Behavioral Development, 18*, 489–504.

Doyle, A. B. (1975). Infant development in day care. *Developmental Psychology, 16*, 31–37.

Dunn, J. (1988). Connections between relationships: Implications of research on mothers and siblings. In R. A. Hinde & J. Stevenson-Hinde (Eds.), *Relations between relationships* (pp. 168–180). Oxford: Clarendon Press.

Dunn, J. (1993). *Young children's close relationships: Beyond attachment.* Newbury Park, CA: Sage.

Eisenberg, N., Fabes, R. A., Schaller, M., Carlo. G., & Miller, P. A. (1991). The relations of parental characteristics and practices to children's vicarious emotional responding. *Child Development, 62*, 1393–1408.

Etaugh, C. (1980). Effects of nonmaternal care on children: Research evidence and popular views. *American Psychologist, 35*, 309–319.

Garner, P. W., Jones, D. C., & Miner, J. L. (1994). Social competence among low-income preschoolers: Emotion socialization practices and social cognitive correlates. *Child Development, 65*, 622–637.

Grolnick, W. S., Bridges, L. J., & Connell, J. P. (1996). Emotion regulation in two-year-olds: Strategies and emotional expression in four contexts. *Child Development, 67*, 928–941.

Harms, T., & Clifford, R. (1980). *Early childhood environment rating scale.* New York: Teachers College Press.

Harms, T., Jacobs, E. V., & White, D. R. (1995). *School age care environment rating scale.* New York: Teachers College Press.

Harris, P. (1994). The child's understanding of emotions: Developmental change and the family environment. *Journal of Child Psychology and Psychiatry, 35*, 3–28.

Hartman, H. I., & Pearce, D. M. (1989). *High skill and low pay: The economics of child care work.* Washington, D.C.: Institute for Women's Policy Research.

Haskins, R. (1985). Public school aggression among children with varying day-care experience. *Child Development, 56*, 689–703.

Hayes, C., Palmer, J., & Zaslow, M. (1990). *Who cares for America's children? Child care policy for the 1990's.* Washington, DC: National Academy Press.

Hestenes, L., Kontos, S., & Bryan, Y. (1993). Children's emotional expression in child care centers varying in quality. *Early Childhood Research Quarterly, 8*, 295–307.

Hofferth, S. L. (1992). The demand for and supply of child care in the 1990's. In A. Booth (Ed.), *Child care in the 1990's: Trends and consequences* (pp. 3–25). Hillsdale, New Jersey: Erlbaum.

Howe, N., & Rinaldi, C. R. (1996). Sibling caretaking: Practical knowledge and the role of maternal emotional socialization. Poster presented at the International Society for the Study of Behavioral Development, Quebec City, CA.

Howe, N., & Rinaldi, C. R. (1997). Sibling caretaking: The role of maternal socialization and sibling-directed internal state discourse. Paper presented at the Society for Research in Child Development, Washington, DC.

Howes, C., & Olenick, M. (1986). Family and child care influences on toddler's compliance. *Child Development, 57*, 202–216.

Howes, C., Rodning, C., Galluzzo, D, & Myers, L. (1988). Attachment and child care: Relationship with mother and caregiver. *Early Childhood Research Quarterly, 3*, 403–416.

Howes, C., Smith, E., & Galinsky, E. (1995). *Florida quality improvement study: Interim report.* New York: Families and Work Institute. Cited in Kontos, S. & Wilcox-Herzog, A. (1997).

Kagan, J., Kearsley, R. B., & Zelazo, P. (1976). The effects of infant day care on psychological development. Paper presented at the American Association for the Advancement of Science, Boston, Mass.

Kamerman, S., & Kahn, A. (1981). *Child care, family benefits, and working parents: A study in comparative policy.* New York: Columbia University Press.

Kontos, S., & Dunn, L. (1993). Caregiver practices and beliefs in child care varying in developmental appropriateness and quality. In S. Reifel (Ed.), *Advances in early education and day care. Vol. 5.* pp. 53–74. Greenwich, CT: JAI press.

Kontos, S., & Wilcox-Herzog, A. (1997). Teachers' interactions with children: Why are they so important? *Young Children, 52*(2), 4–12.

Love, J., Ryer, P., & Faddis, B. (1992). *Caring environments — Program quality in California's publicly funded child development programs: Report on the legislatively mandated 1990–91 staff/child ratio study.* Portsmouth, NH: RMC Research Corporation. Cited in Kontos & Wilcox-Herzog (1997).

McCartney, K. (1984). Effect of day care environment on children's language development. *Developmental Psychology, 20,* 244–260.

Mill, D. (1997). Correlates of affectionate and angry behavior in day care educators of preschool-aged children. Doctoral Thesis, Department of Psychology, Concordia University, Montreal, Quebec.

Phillips, D., Howes, C., & Whitebook, M. (1991). Child care as an adult work environment. *Journal of Social Issues, 47,* 49–70.

Phillips, D., McCartney, K., & Scarr, S. (1987). Child-care quality and child development. *Developmental Psychology, 23,* 537–543.

Phillips, D., Scarr, S., & McCartney, K. (1987). Dimensions and effects of child care quality: The Bermuda study. In D. A. Phillips (Ed.), *Quality in child care: What does research tell us?* (pp. 1–19). Washington, D.C.: National Association for the Education of Young Children.

Ramey, C. T., Bryant, D. M., Campbell, F. A., Sparling, J. J., & Wasik, B. H. (1990). Early intervention for high-risk children: The Carolina early intervention program. *Prevention in Human Services, 7,* 33–57.

Sroufe, L. A., Schork, E., Motti, F., Lawroski, N., & LaFreniere, P. (1984). The role of affect in social competence. In C. E. Izard, J. Kagan, & R. B. Zajonc (Eds.), *Emotion, cognition and behavior* (pp. 289–319). Cambridge: Cambridge University Press.

White, D. R. (1995). Child care services and kindergarten: Selection, quality and continuity. Executive Summary, Project 4774-91/9(688). Ottawa, Ontario: Child Care Initiatives.

White, D. R., Jacobs, E., & Schliecker, E. (1988). Relationship of day care environmental quality and children's social behavior. *Canadian Psychology, 29,* Abstract #668.

White, D. R., & Mill, D. (in press). Current social issues: The child care provider. In L. Prochner & N. Howe (Eds.), *Early childhood education in Canada: Past, present and future.* Vancouver, BC: UBC Press.

Zahn-Waxler, C., Ridgeway, D., Denham, S., Usher, B., & Cole, P. (1993). Research strategies for assessing mothers' interpretations of infants' emotions. In R. Emde, J. Osofsky, & P. Butterfield (Eds.), *Parental perceptions of infant emotions* (pp. 217–236). Madison, Ct: International Universities Press.

6

COMPETENCE[*]

A Short History of the Future of an Idea

William M. Bukowski, Tanya A. Bergevin, Amir G. Sabongui, and
Lisa A. Serbin

Centre for Research in Human Development
Department of Psychology
Concordia University

Competence is one of psychology's most popular expressions and constructs. It appears in the titles of large numbers of books, articles, symposia, and presentations. Like the terms *risk*, *stress* and *vulnerability*, the term *competence* does not refer to a particular or singular process or phenomenon but is instead typically used as a collective term that refers to a general class or category of phenomena regarding behaviour. In this way, the term competence is used as an enveloping term, bringing together or organizing the various components of a research program or a body of scientific literature. That this term should refer to such a broad range of phenomena is useful for psychologists in the sense that it serves a powerful organizational function allowing researchers and theoreticians from diverse areas to see the metatheoretical links that bring them together. In this respect a broad term like competence facilitates communication between persons who maintain a similar attitude toward human growth and functioning.

Nevertheless, there is a cost to the breadth related to the term competence. As the term competence becomes evermore inclusive, its specific meaning becomes increasingly amorphous, elusive, and, ultimately, vague. As this vagueness cuts into the meaning that the term competence conveys, its utility as a psychological construct becomes increasingly limited. At some point it becomes just another buzz word in the lexicon of psychobabble, too weak to facilitate the sort of specificity and exactness that are needed for the development of compelling theory and fruitful empirical investigations of behaviour.

* This paper was written with the support of the Social Sciences and Humanities Research Council of Canada, the Fonds pour la formation de recherche et pour l'aide a la recherche, and the W.T. Grant Foundation. Direct correspondence to the author at Department of Psychology, Concordia University, 7141 rue Sherbrooke Ouest, Montréal, Québec, H4B 1R6, Canada.

Improving Competence across the Lifespan, edited by Pushkar *et al.*
Plenum Press, New York, 1998.

In this chapter we explore the concept of competence so as to identify the nature of its meaning and to show how this concept can be used profitably in research on social development. There are two features to the perspective we adopt. First, we see competence as a condition of the relationship between an individual and the environment. In this regard, one can not think of individuals as being competent without knowing the contexts in which they are engaged. Second, following from this we recognize that the term competence functions as an expression of an attitude toward the study of human growth. By referring to an interactional state between an individual and its environment, the term competence necessarily implies that the individual's competence can be understood only in terms of its environment. That is, it represents an orientation of how one should understand or conceive of human development.

The first part of our discussion of the term competence takes an etymological perspective. In an effort to determine what this word means, we try to see where it came from by identifying its roots. As part of this discussion we point to the various meanings that are ascribed to it. This discussion reveals the specific links between the term competence and other terms and expressions. Second, we examine how the construct of competence was used in one of the most widely cited articles in the history of psychology, namely, Robert White's 1959 *Psychological Review* article "Motivation reconsidered: The concept of competence." We discuss this article according to its content (i.e., the particular arguments that White makes) and the trajectory of its impact on the study of development. An important aspect to this discussion is a consideration of the historical circumstances in psychology in which White's perspective took hold and the reasons why the impact of this article has now waned completely. In the third section of the chapter, we present a conceptualization of what it means to be competent within the peer domain during early adolescence. The model we adopt recognizes that competence can be seen at different levels of social complexity (i.e., the individual, the dyad, and the group) and that in spite of their conceptual independence, these three levels are intricately interrelated.

WHAT DOES THE WORD *COMPETENCE* MEAN?

An Etymological Analysis. An examination of natural language is one way that social scientists can explore the meaning of particular constructs (e.g., see Krappmann's (1996) language-based analysis of the concept of friendship). An analysis of the origins of a word (i.e., an etymological analysis) can reveal how a word was originally formed, thus shedding light on the meaning and power that a word possesses. This sort of approach is especially useful when the exact meaning of a term is not clear or has been allowed to become vague through over-use. Certainly, these characteristics are true for the word *competence*.

The *Oxford English Dictionary* (OED; Simpson & Weiner, 1989) shows that the word *competence* literally[1] follows from the word *compete*. Initially, *compete* represented the union of the words *cum* (i.e., *with* or *together* in Latin) and *petere* (i.e., "to aim for or to go toward"). Literally, then, the term *compete* originally meant to pursue a goal together with or in the company of others. The *OED* points out that compete is a rather recent term, not being found in any of the major lexicons of the 18th and early 19th centuries (e.g., Johnson's) and that it was not until 1824 that it was recognized as a word that was "not uncommon." Competence is described as being a direct derivative of the word *compete*, probably entering the English language from the french word *compétence*, an adjective indicating that one is able to engage in competition. In parallel to this, the word *competent* is shown to be a direct derivative of the present participle of the Latin *competere*, literally meaning "competing."

Two features of these definitions are important. First, the direct link between the words *compete* and *competence* is undeniable. To be competent means that one is able to "compete." It is important to recognize though that the meaning of *compete* to which the meaning of *competent* is linked is not the modern (i.e., late 20th century) meaning of compete which emphasizes the struggle between persons, but is instead the original meaning — that is the striving for mastery. Second, this analysis shows that competence is not a state, but is instead a form of activity. Rather than being an end point of development, competence refers to the engagement of an individual toward the acquisition of a particular state. Therefore a competent person is not a person who possesses a particular skill or ability but is a person who is able to achieve a particular goal. This distinction between competence as a type of action rather than as a state is nontrivial. Indeed these two views represents fundamentally different attitudes toward development, specifically whether the essential goal of development is a product or a process.

ROBERT WHITE — THE FATHER'S GERMINAL PAPER

This exact view — that competence is a process — serves as the centre piece of one of the most influential[2] papers in developmental psychology, specifically Robert White's 1959 article in Psychological Review titled "Motivation considered: The concept of competence." In this paper, White integrates findings and perspectives from several disparate sources — animal behaviour, behavioral neurobiology, neoFreudian psychoanalysis, social psychology, developmental theory about play — to develop an argument that persons are inherently oriented toward an efficient interaction with the environment. He argues that this orientation is one of the most fundamental processes of human behaviour, one that organizes behaviour and organizes development itself. He defines this process — that is, an organism's orientation toward an efficient interaction with the environment — as competence.

For White, this orientation was a profound and basic form of motivation, as critical as any fundamental drive. According to his view, competence is not a form of motivated behaviour that serves other drives, but is a fundamental form of motivation in and of itself. In other words, competence is not a process in the service of other drives or motivations but is a drive and motivation per se. To support this argument, he refers to the high rates of exploratory behaviour seen in many animals and the instinctual curiosity about novelty present in babies. Both of these phenomena show that organisms are inherently oriented toward interaction with the environment. He finds evidence of this "instinct to master" in the observations and interpretations provided by neo-analysts (e.g., Fenichel, 1945; Hendrick, 1942). Hendrick, for example, goes so far as to say that individuals have an "inborn drive to do and to learn to do." Again, White is careful to point out that according to these neoanalysts, such activity is not in the service of another goal (e.g., anxiety reduction), but is instead a basic human motivation by itself.

White (1959) shows also that this view that an orientation toward mastery is central to development is made explicit in the writings of another analyst, namely Erik Erikson. In Erikson's (1953) description of the latency years, he emphasizes that this developmental period is a time in which competence is the main developmental struggle that children confront. Erikson says that "children *need* a sense of being able to make things and to make them well and even perfectly: this is what I call a sense of industry."

White's article (1959) appeared just prior to the time when Piaget's writings were first becoming well-known and understood in North America. Although Piaget's ideas were not yet in wide circulation, White was prescient in his recognition of the power of

Piaget's model. Moreover, he saw the connections between his ideas and the model of adaptation central to Piaget's theory. White recognized that Piaget's belief that the dynamic interaction between an organism and the environment was constantly oriented toward a state of adaptation or equilibrium parallelled his concept of competence. Both he and Piaget emphasized the dynamic quality of development and they believed that this dynamism was directed toward a match between the individual and the environment. In its most basic manifestation, the model of Piaget parallels, in some respects, White's.

The parallels between White and Piaget are probably of independent origin in the sense that neither of them was directly influenced by the other. Nevertheless, the popularity and impact that their perspectives have had are probably due to their shared contrast with the general intellectual climate in psychology during the 1950s. At this time, the predominant model of development was rooted in behaviourism, a perspective that treated the organism as passive rather than active and saw motivation as largely under the control of external forces. According to this view, it was difficult to conceive of competence as a process or as a form of interaction. Instead, from the behaviourist perspective, competence was conceptualized as frequencies of behaviour or as behaviours that occurred in response to environmental circumstances. The position of Piaget and White, specifically that persons were motivated internally to achieve an adaptation or an effective interaction with the environment, directly contradicted this view. Their model of the person as actively involved in his/her development provided a powerful counterpoint to the behaviourist model of the individual as passive. This alternative view turned out to be very attractive to persons concerned with the processes underlying development. To a large extent the perspectives of Piaget and White are still with us today.

The impact of White's views is, in part, apparent in the definitions of competence that subsequent writers have offered. These include the following: "the effectiveness or adequacy with which an individual is capable of responding to various problematic situations which confront him" (Goldfried & D'Zurilla, 1969, p. 161); "an individual's everyday effectiveness in dealing with his environment" (Zigler, 1973); "a judgment by another that an individual has behaved effectively" (McFall, 1982, p. 1); "attainment of relevant social goals in specified social contexts, using appropriate means and resulting in positive developmental outcomes" (Ford, 1982, p. 324); the ability "to make use of environmental and personal resources to achieve a good developmental outcome" (Waters & Sroufe, 1983, p. 81); and "the ability to engage effectively in complex interpersonal interaction and to use and understand people effectively" (Oppenheimer, 1989, p. 45). In a review of these definitions, Rubin, Bukowski, & Parker (in press), point out that they share at least two points: (a) an emphasis on effectiveness, and (b) one understanding of the environment so as to achieve one's own needs or goals. Nevertheless, in spite of these apparent points of similarity between these definitions and that of White, there is one striking difference. In these definitions, competence is treated as an entity or a condition — the end point of a process — or as an ability — rather than as a form of action *per se*. The dynamism that stood at the centre of White's conceptualization appears to have vanished.

There may be a reason for this. Among psychologists, nothing is more challenging to study than action. It is difficult to observe directly, difficult to quantify, and difficult to define. Relative to measuring action, especially internal action, it is easier to measure whether someone has achieved an outcome, performed a task, or obtained a goal. Therefore, we probably should not be too surprised that persons have opted to define competence in terms of outcome rather than in terms of process. In spite of these measurement issues, however, it is not necessary to abandon the central force of White's perspective. Researchers need to recognize that even though we may be forced by our weak methods to measure static events, the static events we ultimately choose to measure should tell us something about process and action. Pursuing this goal is, in our view, evidence of competence in research psychologists.

SOCIAL COMPETENCE IN CHILDHOOD AND ADOLESCENCE

Before deciding how to find competence, one must know where to find it. This concern with "where is competence" has been central to recent advances in the study of peer relations. We discuss these advances and show how they inform us of how competence is manifested in children's and adolescents' experiences within the peer system. In this section we describe an approach to the study of competence that is predicated on multiple levels of analysis. As part of this discussion, we look to findings from research on children's experiences with peers to help us understand the processes underlying competence for school-age and adolescent boys and girls.

Where Is Competence? Persons do not exist apart from others. We hardly ever act alone; for many persons, contact with others is the "ground" of existence. Typically, our actions are part of a larger constellation of behaviours that is defined by our own actions and motivations as well as by those of others. Indeed, persons are embedded in a complex social structure in which they act and behave in conjunction with others.[3] This fundamental interpersonal nature of human beings makes it is easy to address the question "where is competence?" The answer is that competence, or the lack of competence, is most likely to be seen in our experiences with others. This answer is, in our view, true across the life span but is especially true during childhood and early adolescence. Children spend large amounts of time in the company of their peers; the peer environment is the one to which children are most likely to be oriented for efficient interaction. Nevertheless, the answer that the peer system is the locus of competence is somewhat inadequate because the peer environment is so diverse and multifaceted. Instead of being limited to one level of analysis, peer experiences happen in a variety of social and personal contexts.

Consistent with this perspective, it is our view that competence in the peer system is found at many levels of social complexity. The view we adopt bears a resemblance to a comment made by Thoreau in his description of his cabin near Walden Pond. Thoreau said "I have three chairs — one for solitude, two for friendship, and one for society." So too can the peer system be seen as having three levels of social complexity, specifically (a) the individual, (b) the dyad, and (c) the group.

Peer researchers have always been aware that multiple dimensions of experience are subsumed within the peer system. At times however, the distinctions between these dimensions have been blurred. For example, investigators have confused phenomena from different levels (e.g., failing to distinguish between popularity and friendship) or have used a measure from a single level as a general index of a child's experiences with peers. More recently, there has been a heightened concern with differentiating among phenomena from the group, dyadic and individual levels and with understanding how these levels are interrelated. Much of this concern reflects advances in developmental theory regarding the understanding of relationships. Indeed a broad concern with recognizing the multiple levels of experience in which development occurs was a central theme in the writings of Bronfenbrenner and Crouter (1983) and Hinde (1979, 1987, 1995). Our point is that these different levels of complexity provide different contexts for engaging in the processes we would regard as competence.

This multiple level model provides a means of understanding how competence is manifested and it gives us a means of understanding how experiences with peers affect development. One goal of research on peer relations has been to identify competence at different levels and to understand how these levels are related to each other. Regarding this latter point, it should be stated clearly that in spite of the conceptual distinctions between the group, the dyad and the individual, these levels are inextricably interrelated. Phenomena at each level are constrained and influenced by phenomena at other levels. For this exact reason, one of the most important goals of recent research has been to understand the links between group, dyadic and individual constructs.

The Individual. The level of the individual is the most basic level of social complexity. Constructs conceptualized at the level of the individual refer to the properties and characteristics that persons generally bring with them to their activities and experiences (Kramer, Bukowski, & Garvey, 1989). In a sense the level of the individual has no social complexity. Individuals, by themselves, can not engage in any real social action. Accordingly, in so far as competence is a form of action, this level of analysis is limited in what it can tell us about competence. At best, we can identify whether an individual engages in behaviour that we would interpret as indicating competence.

The aspect of behaviour taken from the level of the individual that has been studied most broadly concerns individuals' general social tendencies. In parallel to psychology's general concern with the processes of (a) moving against others, (b) moving toward others, and (c) moving away from others, research on behaviour and experience with peers has been largely organized around the constructs of aggression, sociability and helpfulness, and withdrawal. This interest in how the child's individual behaviours influence experiences with peers has comprised a large portion of research on the peer system during the past 15 years. In fact, the very large literature on children's popularity has been typically focused on behaviours at the level of the individual. Information about the correlates of popularity is important as it reveals which characteristics from the level of the individual are related to effectiveness in establishing a general place within the peer group.

In a meta-analysis of this large body of literature regarding individual behaviours and popularity, Newcomb, Bukowski, and Pattee (1993) were able to reach conclusions about differences between individuals from different popularity groups, especially, the popular, rejected, and neglected groups. They were able to interpret these differences as a function of what they reveal about competence. Specifically, they reported that popular children (i.e., those who were liked by many children and disliked by very few) are more likely to be helpful, to interact actively with other children, to show leadership skills, and to engage in constructive play (cf. Newcomb et al., 1993). These findings show that popular children are capable of pursuing and achieving their own goals while simultaneously allowing others to do the same. In other words, they not only engage in "competent" action themselves, but they are able to promote it in others. One critical finding in regard to the behaviour of popular children is that there are no differences between them and other children on all aspects of aggression. Newcomb et al. differentiated assertive/agonistic behaviors and disruptiveness. Popular children did not differ from others on the dimension of aggression but they were significantly less disruptive. These findings show that although popular children engage in assertive and agonistic behaviour, these assertive actions are unlikely to interfere with the actions and goals of others. Consistent with other reports (Olweus, 1977; Pepler & Rubin, 1991), it would appear as if some forms of aggression may be compatible with being popular and with the concept of social competence. In contrast with the popular children, Newcomb et al. reported that rejected children not only showed high levels of disruptive behaviour but also showed little evidence of engagement in activities with peers. In other words, their activities disrupted their peers' effective engagement with the environment, and showed little effective engagement themselves. This latter point is true of the neglected group also.

A second phenomenon from the level of the individual that has been studied is that of social cognition. Social cognition includes a broad array of processes, including the way that an individual interprets and understands the behaviours of one's self and of others. Akin to research on the link between behaviour and experiences with peers, much of the research on social cognition has examined how an individual's use of social information is linked to the person's experiences with peers at the group level. Indeed, research on social cognition has also examined how an individual's understanding and interpretation of experience may partially determine the impact that experiences with peers have on development.

This sort of social understanding is likely to be a prerequisite for competence. Without knowing how others will act, and without having a means of understanding the motives of others, individuals do not a have a "ground" on which to base their action (Harré, 1974). Moreover, in order to effectively interact with the environment, one needs to possess the "local knowledge" about how this environment functions (Geertz, 1983). Considering this, it is not surprising that most models of social competence place a strong emphasis on social understanding as a powerful antecedent to competent social action (Crick & Dodge, 1994; Dodge, 1986).

The Dyad. Peer experiences at the level of the dyad can be conceptualized as reflecting either interactions and relationships. According to Hinde (e.g., 1979, 1987, 1995) interactions are the series of interchanges that occur between individuals over a limited time span. Interactions are defined as interdependencies in behaviour (Hinde, 1979). That is, two persons are interacting when their behaviours are mutually responsive. If a person addresses someone (e.g., says "Have a nice day") and the other person responds ("Don't tell me what to do") they have had an interaction, albeit a rather primitive one. Relationships are based on these patterns of interaction. The relationship consists of the cognitions, emotions, and internalized expectations and qualifications that the relationship partners construct as a result of their interactions with each other.

Hinde (1979, 1995) has shown researchers can examine relations according to (a) what the relationship partners do together (i.e., the content of the relationship); (b) the number of different activities the partners engage in (i.e., the breadth of their interactions); and (c) the quality of their interactions within a relationship (e.g., reciprocally, complementarily, positively, negatively). As in the research on popularity in which associations between behaviour and popularity were used as the basis of conclusions regarding competence, research on friendship, a primary form of relationship for children, informs us of the processes that are central to engagement at the dyadic level of experience. In research on friendship, different characteristics of relationships can be contrasted empirically, for example, between friends and acquaintances or disliked others.

One domain of interaction which nicely illustrates the nature of competence in friendship relations is that of conflict. Conflict is one of the few dimensions of interaction in which there are no differences between friends and nonfriends. Research has shown repeatedly that during childhood and adolescence, pairs of friends engage in the same amount of conflict as pairs of nonfriends (Hartup, 1992; Laursen, Hartup, & Koplas, 1996; Newcomb & Bagwell, 1995). There is, however, a major difference in the conflict resolution strategies that friends and nonfriends adopt. In a large meta-analysis of research on friendship, Newcomb & Bagwell showed that friends were more concerned about achieving an equitable resolution to conflicts than achieving outcomes that would favour one partner. More specifically, Laursen et al. and Hartup reported that friends are more likely than nonfriends to resolve conflicts in a way that will preserve or promote the continuity of their relationship. In this respect competence in friendship appears to require the ability of the two partners to find a balance between individual and communal goals.

A significant feature of this interpretation of these findings is its implicit recognition that competence in friendship is based on achieving a balance between personal desires against social consequences. This emphasis reflects an essential duality of self and other, placing the individual within a social and personal context. The conceptualization of self and other as interdependent is an important feature of theory from several domains including personality theory (Bakan, 1966; Sullivan, 1953), gender role theory (Leaper, 1994), feminist psychology (Jordan, Kaplan, Miller, Steiver, & Surrey, 1991) and philosophy (Harré, 1979, 1984). Accordingly, this definition is valuable by pointing to the complex goals that persons confront as individuals and as members of groups. It is in this regard that we believe that competence in friendship involves resolving the tension between agentic and communal orientations.

The Group. Group experiences refer to experiences among a set of individuals who have been organized by either a formal or an informal means. Group phenomena do not refer to individuals but instead refer to (a) the links that exist between the persons in the group, and (b) the features that characterize the group. Groups can be measured according to (a) their structure and size (cf., Bennenson, 1990), and (b) the themes or the content around which groups are organized (cf., Brown, 1989). Many of the best known studies of children and their peers were concerned with the group *per se*, including Lewin, Lippit and White's (1938) study of group climate, and Sherif, Harvey, White, hood, & Sherif's (1961) study of intragroup loyalty and intergroup conflict.

In spite of the apparent importance of phenomena at the level of the group, group measures have received far less attention than measures conceptualized at the level of the dyad and the individual. This is surprising because persons generally refer to the peer system as a group phenomenon. Indeed references to experiences with peers invariably refer to the "peer group." Moreover, as Hartup (1983) noted, early theory and research on the peer system was organized around the idea that group experiences were powerful determinants of behaviour. Nevertheless, research on the factors related to the peer group *per se* has been relatively sparse. The reasons for this neglect of the group can be traced to the complexity of the conceptual and methodological issues related to the study of the structure and organization of the group. In particular, questions regarding the basic properties of groups and how these properties should be measured can appear to be more challenging than issues related to characteristics of individuals and dyad. Accordingly, persons have devoted their research resources to other levels of analysis.

In spite of a relative lack of research, the studies regarding peer group behaviour generally converge on the same conclusion. This conclusion is that groups that show the most effective engagement with the environment (e.g., those that can complete a task most efficiently) are those that have the strongest sense of cohesion and shared purpose (e.g., Sherif et al., 1961) or that have the most engaged leadership (Lewin, Lippet, & White, 1938). These studies show that one of the factors that appears to precede competence for an individual is critical also for that of a peer group. This factor is the shared knowledge or understanding of a task or ritual necessary for a group to function toward a common goal.

SUMMARY

Our emphasis on multiple levels of analysis provides more than just a means of thinking about the peer system. It also gives us a basic model for developing an idea of what constitutes social competence. Implicit in the perspective we have taken is the argument that social competence within the peer system refers to a child's capacity to engage effectively and successfully at each level of analysis. By this we mean that a competent child will be able to (a) become engaged in a group structure and to participate in group-oriented activities, (b) become involved in satisfying relationships constructed upon balanced and reciprocal interactions, and (c) satisfy individual goals and needs and develop accurate and productive means of understanding experiences with peers on both the group and dyadic levels. We see two particular processes as being central to competence at each of the levels, namely (a) the capacity to balance individual goals with those of others, and (b) the ability to understand the social goals and actions of others.

In our discussion of social competence we have argued that competence is a form of social action. For us, social competence refers to an effective engagement in interactions with others. In this regard, competence is not a "thing" but is a process in which persons move toward goals that are personal as well as interpersonal, achieving their own objectives as well as allowing others to achieve their goals. Studying competence as a form of action presents a challenge to psychologists who are more accustomed to studying phenomena that can be measured as entities or quantities. We, as researchers and persons interested in enhancing competence in persons across the life-span would be served well by addressing this challenge. This is a goal we can pursue together in the company of others, which, as we note at the beginning of the chapter, is exactly what competence means.

ENDNOTES

1. By this we mean that the word competence and derivatives of it come directly after the word compete in the *OED*. Interestingly, these words are actually sandwiched between "compete and "competition."
2. White's article was widely cited in every year since its publication. Across the sixties it was cited; during the '70s, it was cited with increasingly frequency every year, starting with 23 citations in 1970 and finishing with at 67 citations in 1979. In 1980, 1981, and 1982 it was cited 80 times, 83 times, and 72 times, respectively, and was cited at least 43 times in each year until the present.
3. Even cognition is a deeply interpersonal phenomenon (see Hartup, 1996).

REFERENCES

Bakan, D. (1966). *The duality of human existence*. Boston: Beacon Press.

Benenson, J. F. (1990). Gender differences in social networks. *Journal of Early Adolescence, 10*, 472–495.

Bronfenbrenner, U., & Crouter, A. C. (1983). The evolution of environmental models in developmental research. In E. M. Hetherington (Ed.), *Handbook of child psychology: Vol. 1. History, theories and methods* (4th ed., pp. 357–414). New York: Wiley.

Brown, B. B. (1989). The role of peer groups in adolescents' adjustment to secondary school. In T. J. Berndt & G. W. Ladd (Eds.), *Peer relationships in child development* (pp. 188–216). New York: Wiley.

Crick, N. R., & Dodge, K. A. (1994). A review and reformulation of social information processing mechanisms in children's social adjustment. *Psychological Bulletin, 115*, 74–101.

Dodge, K. A. (1986). A social information processing model of social competence in children. In M. Perlmutter (Ed.), *Minnesota symposium on child psychology: Vol. 18* (pp. 77–125). Hillsdale, NJ: Erlbaum.

Erikson, E. H. (1953). Growth and crises of the healthy personality. In C. Kluckhorn, H. A. Murray, & D. Schneider (Eds.), *Personality in nature, society, and culture* (pp. 185–225). New York: Knopf.

Fenichel, O. (1945). *The psychoanalytic theory of neurosis*. New York: Norton.

Ford, M. E. (1982). Social cognition and social competence in adolescence. *Developmental Psychology, 18*, 323–340.

Geertz, C. (1983). *Local knowledge: Further essays in interpretive anthropology*. New York: Basic Books.

Goldfried, M. R., & D'Zurilla, T. J. (1969). A behaviour analytic model for assessing competence. In C. D. Spielberger (Ed.), *Current issues in clinical and community psychology* (pp. 151–198). New York: Academic.

Harré, R. (1974). The conditions for a social psychology of childhood. In M. P. M. Richards (Ed.), *The integration of a child into a social world*. London: Cambridge University Press.

Harré, R. (1979). *Social Being*. Cambridge, MA: Harvard University Press.

Harré, R. (1984). *Personal Being*. Cambridge, MA: Harvard University Press.

Hartup, W. W. (1983). Peer relations. In E. M. Hetherington (Ed.), *Handbook of child psychology: Vol. 4. Socialization, personality and social development* (4th ed., pp. 103–196). New York: Wiley.

Hartup, W. W. (1992). Conflict and friendship relations. In C. U. Shantz & W. W. Hartup (Eds.), *Conflict in child and adolescent development* (pp. 185–215). Cambridge, England: Cambridge University Press.

Hartup, W. W. (1996). Cooperation, close relationships, and cognitive development. In W. M. Bukowski, A. F. Newcomb, & W. W. Hartup (Eds.), *The company they keep: Friendship in childhood and adolescence* (pp. 213–237). New York: Cambridge.

Hendrick, I. (1942). Work and the pleasure principle. *Psychoanalytic Quarterly, 11*, 33–58.

Hinde, R. A. (1979). *Towards understanding relationships*. London: Academic Press.

Hinde, R. A. (1987). *Individuals, relationships and culture*. Cambridge: Cambridge University press.

Hinde, R. A. (1995). A suggested structure for a science of relationships. *Personal Relationships, 2*, 1–15.

Jordan, J., Kaplan, A., Miller, J. B., Steiver, I., & Surrey, J. (1991). *Women's growth in connection: Writings from the Stone Center*. New York: Guilford.

Kramer, T. L., Bukowski, W. M., & Garvey, C. J. (1989). The influence of the dyadic context on the conversational and linguistic behaviour of its members. *Merrill-Palmer Quarterly, 35*, 327–342.

Krappmann, L. (1996). Amicitia, drujba, shin-yu, filia, freundschaft, friendship: On the cultural diversity of a human relationship. In W. M. Bukowski, A. F. Newcomb, & W. W. Hartup (Eds.), *The company they keep: Friendship in childhood and adolescence* (pp. 19–40). New York: Cambridge.

Laursen, B., Hartup, W. W., & Koplas, A. L. (1996). Toward understanding peer conflict. *Merrill-Palmer Quarterly, 42*, 76–102.

Leaper, C. (1994). Exploring the consequences of gender segregation on social relationships. In C. Leaper (Ed.), *Childhood gender segregation: Causes and consequences* (pp. 67–86). San Francisco: Jossey-Bass.

Lewin, K., Lippit, R., & White, R. K. (1938). Patterns of aggressive behaviour in experimentally created "social climates." *Journal of Social Psychology, 10*, 271–299.

McFall, R. M. (1982). A review and reformulation of the concept of social skills. *Behavioral Assessment, 4*, 1–33.

Newcomb, A. F., & Bagwell, C. (1995). Children's friendship relations: A meta-analytic review. *Psychological Bulletin, 117*, 306–347.

Newcomb, A. F., Bukowski W. M., & Pattee, L. (1993). Children's peer relations: A meta-analytic review of popular, rejected, neglected, controversial, and average sociometric status. *Psychological Bulletin, 113*, 99–128.

Olweus, D. (1977). Aggression and peer acceptance in adolescent boys: Two short-term longitudinal studies of ratings. *Child Development, 48*, 1301–1313.

Oppenheimer, L. (1989). The nature of social action: Social competence versus social conformism. In B. Schneider, G. Attili, J. Nadel, & R. Weissberg (Eds.), *Social competence in developmental perspective* (pp. 40–70). Dordrecht, The Netherlands: Klure International.

Pepler, D. J., & Rubin, K. H. (Eds.) (1986). *The development and treatment of childhood aggression*. Hillsdale, NJ: Erlbaum.

Rubin, K. H., Bukowski, W. M., & Parker, J. G. (in press). The peer system: Interactions, relationships and groups. In W. Damon (Series Ed.) & N. Eisenberg (Vol. Ed.), *The handbook of child psychology*. New York: Wiley.

Sherif, M., Harvey, O. J., White, B. J., Hood, W. R., & Sherif, C. W. (1961). *Inter-group conflict and cooperation: The Robbers Cave experiment*. Norman, OK: University of Oklahoma Press.

Simpson, J. A., & Weiner, E. S. C. (1989). *The Oxford English dictionary* (2nd ed.). Toronto: Oxford.

Sullivan, H. S. (1953). *The interpersonal theory of psychiatry*. New York: Norton.

Waters, E., & Sroufe, L. A. (1983). Social competence as a developmental construct. *Developmental Review, 3*, 79–97.

White, R. (1959). Motivation reconsidered: The concept of competence. *Psychological Review, 66*, 297–333.

Zigler, E. (1973). Project Head Start: Success or failure? *Learning, 1*, 43–47.

7

ADOLESCENT USE AND ABUSE OF ALCOHOL

Frances E. Aboud and Shelley C. Dennis

McGill University

Alcohol is used and abused across the lifespan. Adolescence is an interesting developmental stage at which to examine alcohol consumption because it is the time when our young become socialized to drink with control and moderation in a social context. Different cultures around the world and in North America, specifically, use alcohol in somewhat different ways, and adolescents must learn these ways from adults. However, adolescence is also a time when youth assert their independence, and they often use alcohol to experiment with behavior that is counter to adult norms (Fillmore, 1988). This may partly explain why the amount of alcohol consumed is highest in the late adolescent years. It is influenced less by adult norms and more by adolescent norms, which condone experimenting with large quantities of alcohol and disinhibited behavior. In many respects, then, it is quite different from adult abuse of alcohol which tends to manifest itself as alcohol dependency and addiction.

There are many reasons why it is important to examine adolescent drinking behavior. One is to determine whether the patterns of drinking at this age are carried on through adulthood. Specifically, we review research that asks, Are those with a drinking problem in adolescence more likely to have a drinking problem in adulthood? Another purpose of research is to examine the problems that follow from excessive drinking, problems that impact on the drinker, the drinker's family, and society at large. Even if adolescents change their drinking habits when adopting adult roles, they may have to live for many years with the damage alcohol inflicted on their lives. It has been found, for example, that alcohol-related problems are experienced more by minority cultures in North America, even if they drink less. To develop programs that prevent problem drinking before it occurs, those who work with adolescents need to know what conditions place certain adolescents at risk for these problems and how to change the conditions.

Alcohol is still the most widely used drug, legal or illegal, in North America. The consequences of drinking alcohol may be less severe that those that arise from the use of illicit drugs, but they are much more common and thus constitute a more serious social problem. Alcohol has been linked to the loss of many lives both through direct and indirect causes. Eliany (1992) reported that 69% of young Canadians said they had experienced a problem as a result of other people's drinking behavior. Of the 69%, 18% were pushed, hit or physically assaulted, 12% had experienced family problems, and 8% had

Improving Competence across the Lifespan, edited by Pushkar *et al.*
Plenum Press, New York, 1998.

their property vandalized by someone who had been drinking (see also Murdoch, Phil, & Ross, 1990). Almost 3% had been in a car accident caused by someone else's drinking while 23% had been passengers in a motor vehicle with an intoxicated driver. These figures include only those who survived the assault or motor vehicle accident. Estimates in the United States place the percent of motor vehicle deaths that are due to intoxicated drivers at 42% (Hurley & Horowitz, 1990). If they survive the crash, adolescents may sustain injuries that disable them for many years. Excessive drinking also leads to other serious problems, such as family conflicts and poor school performance. Legal and school problems, in particular, can have a lasting effect on the life of an adolescent at a time when they are building competencies and credentials for their future.

This chapter will examine adolescent drinking by looking firstly at its definition and measurement, and secondly at the prevalence of alcohol use and abuse of White and minority adolescents. Finally we will examine risk factors that identify which adolescents are likely to have alcohol problems, and the preventive interventions that attempt to reduce alcohol abuse by minimizing these risks.

DEFINITION AND MEASUREMENT

There have been many attempts to reach a clear definition of alcoholism which can be widely accepted. Aronson (1987) concluded that alcohol abuse is on a continuum, that includes normal drinking patterns, because there is no definite threshold separating problem drinking from normal drinking. However, in practice, clinicians using the American Psychiatric Association's *Diagnostic and Statistical Manual* (DSM-IV; 1994) have identified two forms of alcoholism: alcohol abuse and alcohol dependence. Alcohol abuse is the broader term referring to a compulsive and recurrent use of alcohol resulting in social, psychological, or physical problems. One or more of the following alcohol-related problems are experienced within a 12-month period: (a) failure to fulfill major role obligations at work, school, or home, such as school suspensions, absences, poor performance; (b) recurrent use in physically hazardous situations such as driving an automobile; (c) legal problems such as arrests; and (d) continued use despite persistent social or interpersonal problems exacerbated by the effects of alcohol, such as fights with parents and friends. Dependency specifically refers to long-term abuse of alcohol leading to three or more out of seven problems: tolerance, withdrawal symptoms, uncontrolled drinking, a preoccupation with finding alcohol, and so on. The term dependency does not usually apply to adolescents because they have not abused alcohol for a sufficiently long time to experience tolerance, withdrawal or the other symptoms. However, the current definition of alcohol abuse is particularly relevant because it places emphasis on the social and psychological problems that arise from excessive drinking, rather than on the addiction to alcohol per se. It is therefore relevant to adolescent abuse of alcohol as well as to adult abuse, and it is flexible enough to accommodate the norms of different cultures.

Instruments that measure alcohol abuse parallel its current definition. Large-scale surveys of households or of high school students generally include questions about the frequency of drinking, the number of drinks per occasion, and alcohol-related problems (see Oetting & Beauvais, 1990). The frequency question may be: How often do you drink in a typical month? The quantity question is: How many drinks do you have on a typical occasion? By multiplying frequency by quantity, they arrive at the number of drinks consumed in a typical month. To determine the prevalence of episodic heavy drinking, or binge drinking, they may ask if the person ever consumes six or more drinks at one sitting. Adolescents in North America usually drink beer which contains 3–5% pure alcohol, but in places where both beer and stronger spirits are consumed, the amounts may be translated into quantities of pure alcohol (Miller, Heather, & Hall, 1991).

Lifetime use in these surveys refers to ever having consumed alcohol. In contrast, current users of alcohol are defined as those who have consumed alcohol at least once in the past 12 months. A finer breakdown of users may be made to distinguish Light and Moderate from Heavy users (those who drink at least once a week and typically have five or more drinks on each occasion; Wilsnack & Wilsnack, 1978). Gower (1990) defined "heavy drinkers" as women who reported drinking more than 11 drinks and men who drank more than 14 drinks a week or people who consumed five or more drinks on one occasion 52 or more times in the past year. Alcohol-related problems include the ones mentioned previously, such as driving while intoxicated, violence, suspension from school and absenteeism, and family conflicts.

PREVALENCE OF USE AND ABUSE

There are several major surveys conducted regularly in the United States on alcohol use — the National Senior Survey of final year high school students, the Drug and Alcohol survey of younger students, and the National Household Survey targeting people of all ages. In Canada, the Addiction Research Foundation and Statistics Canada conduct surveys of alcohol use. To put adolescent alcohol use in a lifespan context, we begin by including some statistics on use in the larger society. The National Household Survey reported that 83.6% of Americans aged 12 years and over had at sometime consumed alcohol, though only 49.6% were current users (Giovino, Henningfield, Tomar, Escobedo, & Slade, 1995). In Canada, 55% of the population 15 years and over had consumed alcohol in the past month (Millar, 1991). In both Canada and the United States, rates of drinking have declined since the early 1980s. The lifetime prevalence of alcohol abuse is close to 20% in both countries, though the prevalence of current abuse is less than 10% (Compton, Helzer, Hwu, Yeh, McEvoy, Tipp, & Spitznagel, 1991; Helzer, Canino, Yeh, Bland, Lee, Howes, & Newman, 1990; Hurley & Horowitz, 1990).

In general, one-quarter of the high school students participating in American surveys consumed alcohol in the past month, and half had at some time tasted alcohol (e.g., Glantz & Pickens, 1992). Considering the slightly older population of those 18 to 25 years, 63% had consumed in the past month and close to 90% had tasted alcohol. According to the frequency and quantity indicators, American adolescents between 15 and 19 years averaged 1.4 drinking times per month and 2.8 drinks per occasion (Fillmore, Hartka, Johnstone, Leino, Motoyoshi, & Temple, 1991). In Canada, 47% of males aged 15–19 and 38% of females were current drinkers, meaning drinking at least once in the past month (Millar, 1991).

One out of every four current drinkers (23%) aged 15–24 reported having experienced an alcohol-related problem in the previous year: most were physical health problems (11%), followed by problems with friends or social life (9%), happiness or outlook (6%), work or studies (5%) and home life (5%; Eliany, 1992). Young men (28%) were more likely than young women (18%) to report alcohol-related problems. Binge drinkers, namely those who consumed five or more drinks on 15 or more occasions, were six times more likely to experience problems than other drinkers (48% vs 8%; Eliany, 1992). In 1989, 26% of adolescent males and 14% of adolescent females reported drinking and driving (Eliany, 1992). These general figures indicate that drinking and alcohol-related problems are prevalent in North American adolescents.

Oetting and Beauvais (1990) raise the question of validity of self-reports. Is there any evidence that adolescents are accurately reporting how much and how frequently they drink alcohol, given that most data are of the self-report nature. Their answer is affirmative; self-reports of frequency, namely whether or not one drank in the previous 24 hours, were confirmed with blood alcohol concentrations (BAC). There was a slight bias toward overreporting. Only older alcoholics show a slight bias toward underreporting in comparison with BAC. Surveys on large samples of adolescents are therefore providing valid statistics on alcohol use.

Age Differences

The National Senior Survey conducted since the mid-1970s has consistently found that over 90% of adolescents aged 18 or so have tried alcohol, and that somewhere between 80% and 85% have consumed alcohol in the past 12 months (Glantz & Pickens, 1992; Holden, Moncher, & Schinke, 1990; Johnston, 1985). Johnston claims that heavy party drinking is on the rise at this age; this is supported by the 32.2% who report at least one occasion of heavy drinking in the past 2 weeks (Glantz & Pickens, 1992). Surveys conducted with younger students demonstrate that drinking begins at a young age for some children. Some 22.8% of fourth graders had tasted alcohol (3.3% said they got drunk), but the figure jumps significantly at seventh grade when 65.8% reported drinking at some time and 20% said they got drunk (Oetting & Beauvais, 1990). The average age for the initiation of alcohol use is 14 years.

Longitudinal data on alcohol use indicate that more people use and abuse alcohol during the period between 18 and 20 years, and that the monthly frequency of drinking rises greatly in the years leading up to this period. These conclusions are based on a meta-analysis covering 39 longitudinal studies conducted over the past 30 years (Fillmore et al., 1991). A New York State study followed tenth and eleventh grade students (Kandel & Logan, 1984; Kandel & Yamaguchi, 1985). Initially the researchers found that 51% drank at least four times per week, 24% drank daily, and 50% averaged five drinks per occasion, but 10 years later many had discontinued use. Similarly after 20 years of age, many of the men and women in the Senior Survey had cut their drinking dramatically (Bachman, Wadsworth, O'Malley, Johnston, & Schulenber, 1997). In summary, the number of current drinkers and the frequency of drinking appear to decline after the 18–20 peak. Thus, adolescence is a period when many begin to experiment with alcohol, drinking large quantities frequently. The post-20 decline is thought to reflect a young adult phase of development, when in anticipation of new adult roles, young people curtail their drinking parties.

Cohort (Historical) Differences

The National Senior Survey has shown declines in the use of alcohol since the mid-1980s. Thus, the 1970 cohorts had the highest rates of alcohol and drug use and abuse (Oetting & Beauvais, 1990; Robins & Przybeck, 1985). The proportion of young people consuming alcohol and the amount of alcohol has been dropping since the 1970s. For example, the percentage of high school seniors who drank five or more drinks on one occasion in the past two weeks dropped from 37% to 32% in the late 1980s (Glantz & Pickens, 1992; Holden, Moncher, & Schinke, 1990). In the general population, too, the use of alcohol has been declining. In 1978, the percentage of current drinkers aged 15 and over who consumed 14 drinks or more a week was 20%, in 1985 it dropped to 12%, and declined still further to 11% by 1991. In 1978, 22% consumed 7 to 13 drinks a week which decreased to 18% for 1985 and 1991. In 1978, 42% drank one to six drinks a week which increased in 1985 to 47% but then decreased to 41% in 1991 (Millar, 1991). In 1989, 43% of young Canadians aged 15–24 drank at least once per week on average (Eliany, 1992).

Declines in the use and abuse of alcohol since the 1970s has been attributed to a number of factors. They could be grouped under the theme "promotion of healthy lifestyles" and they seem to have had an impact not only on adolescents but on their parents and the society at large. Education in schools and parent advocacy groups are emphasizing the destructive consequences of alcohol. Parents and Students Against Drinking and Driving are just some of the groups that have been created as a backlash against the number of motor vehicle fatalities resulting from intoxication.

Ethnic and Gender Differences

The 1986 National Senior Survey showed the percentage of lifetime prevalence for White and Black teens. Of a sample of 11,713 White teenagers, 93% had consumed alcohol as opposed to 83% of 1,649 Black teenagers. The 1988 American Drug and Alcohol Survey also made comparisons of different ethnicities for high school samples from the seventh to the twelfth grades. Again a high percentage of 94% was found for the White sample compared to 91% for Reservation Native Indians, 91% for Western Spanish, and 84% for Mexican Americans (Oetting & Beauvais, 1990). These figures are all quite high and have been confirmed by the results of surveys with younger students (e.g., Murray, Perry, O'Connell, & Schmid, 1987; Wilsnack & Wilsnack, 1978).

The Hispanic community is the fastest growing and youngest minority population, displaying the largest number of arrests for drunkeness (138/100,000 population) compared to non-Hispanics (38/100,000; Holden et al., 1990). Mexican-American youth who use alcohol do not get drunk more frequently than their White counterparts, but they do have more problems as a result of drinking. For example, those who drink are more likely to have lower grades, higher drop-out rates, and higher unemployment as adults (Freeman, 1990).

Native Indian teens also have high rates of alcohol use and abuse. The American Drug and Alcohol Survey from 1988 found that 91% of Indian reservation high school seniors used alcohol and 70% got drunk from drinking large quantities (Oetting & Beauvais, 1990). Most researchers and clinicians make the point that although rates of use do not differ from the White adolescent samples, the number of alcohol-related problems experienced by Indian youth are much greater. For example, they are more likely to have conflicts with family members over their drinking habits, more fighting with friends, more hang-overs at school, and more concerns over their reputation with peers (O'Nell & Mitchell, 1996). Alcohol is involved in 90% of all homicides among Indians and 80% of suicides (Freeman, 1990). Because the prevalence of alcohol abuse is high among Indian adults, most youth are exposed to high levels of alcohol use among family members and peers. In one survey, 43% of fathers and 30% of mothers reported a lifetime history of alcohol dependence and 28% had received treatment for alcohol abuse (Walker, Lambert, Walker, & Kivlahan, 1993). This is more than double the rate in the general population. For these reasons, drinking among Indian youth begins by eighth grade and it is seen as a conventional, yet risky, pasttime (Moncher, Holden, & Trimble, 1990; O'Nell & Mitchell, 1996).

Black and Asian males and females showed the lowest prevalence rates of all the ethnic groups (Forney, Forney, Davis, Van Hoose, Cafferty, & Allen, 1984; Holden et al., 1990). Brunswick, Messeri, and Aidala (1990) conducted a longitudinal study of Black teenagers in Central Harlem and found that the average age for initiation of drinking was 14.7 years for boys and 15.3 years for girls. By 18 years, 96% of males and 92% of females had used alcohol. They differed from White samples reported previously in that many of them continued to drink throughout their 20s when 77% of men and 60% of women were current users. Becoming married and having children did not reduce the level of drinking. Blacks who continue drinking into adulthood have a much greater likelihood of dying from cirrhosis than Whites (Freeman, 1990). Thus, although the age of initiating drinking is later and the prevalence of drinking is lower among Blacks generally, among those in Harlem it persisted longer and is likely to lead to more severe medical and social consequences.

Around the world the ratio of men to women alcoholics is 5:1. In Latin American countries the ratio is more like 10:1 and many women abstain from drinking altogether. In North America the ratio is 2:1. Among adolescents, the number of female seniors who drink is consistently 10% lower than the number of males; a larger number abstain from drinking. On the two common indicators of alcohol abuse — quantity-frequency and alcohol-related problems — girls score lower than boys (Wilsnack & Wilsnack, 1978). Girls are also likely to begin drinking a year or two after boys, on average.

Summary

It is clear that the abuse of alcohol is not solely an adult problem. Adolescent drinking, however, is seen as a form of experimentation and intended nonconformity rather than the beginning of a lifelong pattern of alcoholism. The problem among young people is not that they have become dependent on alcohol through prolonged use, but that they abuse alcohol by drinking it in large quantities, frequently, and in many settings outside the home and outside the traditional norms which have in the past controlled drinking.

Recent data have indicated that sometime during the adolescent and young adult years, 35% of drinkers considerably reduce their intake; both the quantity and frequency of drinking decline (Fillmore et al., 1991). However even if drinking declines with age, the damage may have been done if the adolescent drops out of high school, or gets a police record. Thus, the major concern when planning interventions is with problems associated with heavy drinking, such as motor vehicle injuries, violence, delinquency, being expelled from school or fired from a job, and conflicts with parents.

CORRELATES AND CAUSES OF ALCOHOL ABUSE

The study of risk factors for abuse of alcohol attempts to identify correlates of abuse for two purposes. One is to identify adolescents who might be targeted for preventive programs before their drinking becomes more serious; the other is to identify possible causes of alcoholism. In addition to the three demographic variables discussed so far, namely age, gender, and ethnicity, three major categories of variables have been identified (Fillmore, 1988). One concerns personal characteristics of the adolescent, a second concerns the peer and family social environment, and a third concerns access and exposure to alcohol (covered in the Prevention section). Because not all adolescents who use alcohol cross the threshold to abuse, it is important to distinguish users who move to abuse from those who remain users.

With respect to the adolecent drinker, we might ask what is it about age, sex, and ethnic background that make him/her vulnerable to drinking out of control? Is personality or psychopathology relevant to vulnerability? Is there a biological marker for alcohol vulnerability? With respect to the social environment, researchers have examined whether a certain living arrangement or peer norms encourage abuse. Thirdly, is beer too available, too potent, and too inexpensive for people of this age? This will be covered in the last section on prevention.

Correlational studies demonstrate that certain factors are associated with problem drinking. This helps to target high-risk groups of adolescents for preventive programs. However, it is not clear whether the factors are predisposing causal factors, consequences, or a third variable that is only indirectly related to alcohol abuse. Therefore, the best design to identify causes of problem drinking is one that examines people before they become problem drinkers using a prospective approach. Certain studies do this through a longitudinal design, following up adolescents for several years (e.g., Fillmore et al., 1991). Others identify high-risk youngsters, namely sons of alcoholics, who may show predisposing characteristics even during preadolescence (Pihl & Peterson, 1992). This design pulls for genetic, family, and personality predisposing variables. A more inclusive design would be one that follows children who manifest the early predisposing variables through adolescence to see whether they adopt problem drinking patterns, and whether other environmental variables such as peer pressure and beer availability increase the likelihood of abuse.

Interactions between the individual and the environment may take several forms: they may make independent contributions that are additive or interactive. Or the various risk factors may correlate with each other, in that predisposed individuals seek out friends and drinking situations that contribute to the risk of abuse. They may also correlate if there is a "cumulative continuity" of constraining situations brought on by problem behavior; for example, aggressive behavior leads to peer rejection which leads to having only aggressive friends which limits the social activities and skills available to you (Caspi, Elder, & Bem, 1987).

Regardless of whether the risk factors are based in the adolescent or in the environment, findings support the idea that risks are additive — those with only one risk had an 8% chance of abusing alcohol compared to those with four risk factors who had a 43% chance (Brook, Cohen, Whiteman, & Gordon, 1992). Newcomb, Maddahian, and Bentler (1986) followed up White and non-White high school students from grade 7. They found that heavy drinking five years later was predicted by the number of risk factors, five or more being a critical threshold. They included risk factors such as: poor academic achievement, low religious commitment, early alcohol use, poor self-esteem, depression, lack of law abidance, and sensation seeking. Even with many risk factors, half the sample did not abuse alcohol or other drugs. Explaining their resiliency has motivated a search for protective factors (DeWit, Silverman, Goodstadt, & Stoduto, 1995).

Demographic Characteristics of the Adolescent

Age has already been discussed in terms of cross-sectional and longitudinal differences in drinking patterns and alcohol abuse. However, the age at which a person first used alcohol is sometimes considered as a risk factor. The earlier one starts to drink, the more likely it will become a problem sometime in one's lifetime and the more likely it is associated with other behavioral problems such as hyperactivity and conduct disorder (Robins & Przybeck, 1985). However, longitudinal data suggest that having a drinking problem sometime during adolescence is not a good predictor of alcohol problems later in life as most will discontinue heavy use in adulthood (Fillmore et al., 1991). Thus, an early adolescent drinking problem indicates a more serious cluster of current and future problems than a later one.

Gender is a clear risk factor in alcohol use and in problem drinking. Men start drinking at a younger age than women, generally drink larger quantities, and are more likely to have drinking problems later in life. A number of women abstain or drink very lightly. Although the male:female adult ratio for heavy drinking or alcoholism in North America is 2:1, the ratio is closer to 10:1 in many East Asian and Latin American countries. Among adolescents, however, alcohol use and heavy drinking show less of a gender distinction, except among Mexican Americans (Oetting & Beauvais, 1990).

Race and ethnicity are relevant predictors for alcohol abuse in the adolescent years. White adolescents along with Native Americans living on reservations have the highest rates, Mexican Americans the next highest, and African Americans and Asian Americans the lowest. This is surprising to some who tend to assume that all minorities have higher rates of the common social ills. But a similar ranking is found for abuse of other substances such as cocaine, stimulants, and cigarettes (Oetting & Beauvais, 1990). Large scale representative national surveys in the United States consistently yield these figures (e.g., National Household Survey on Drug Abuse; Clayton, 1992). However, smaller scale studies of younger adolescents may show a different pattern. For example, Murray et al. (1987) found the highest rate of heavy drinking among Hispanic (Mexican American) seventh graders, followed by Native American, then White and African American, and the lowest rate among Asian American seventh graders.

Personal Characteristics of the Adolescent

Are problem drinkers also problem adolescents? Correlational studies, to be reviewed here, show a certain amount of co-morbidity and problem clustering. Studies of sons of alcoholic men, who are not yet drinking but are at risk, show certain personality predispositions to alcohol abuse. There appear to be two profiles of problem adolescents who abuse alcohol. One is the hyperactive or aggressive youth with a history of alcohol in the family. Another is the rebellious risk-taker who wants to flout convention by drinking too much and engaging in other disinhibited and stimulating activities.

Among the general population, problems with alcohol are associated with other psychiatric disorders, specifically antisocial personality, depression, and anxiety (Walker et al., 1993). With adolescents, the associations are not always as strong. However, severe abuse of alcohol is most often associated with antisocial personality, labelled conduct disorder in children and adolescents (e.g., Walker et al., 1993). Conduct disorder is manifested as aggression towards people or animals, destruction of property, deceitfulness or theft, or serious violation of rules (DSM-IV). Those who start drinking early are more likely as children to fit this description, namely to have been involved in stealing, frequent lying, running away from home, school expulsion, under-achievement, vandalism, or problem fighting (Robins & Przybeck, 1985). Similarly, Donovan and Jessor (1985) created a composite index of antisocial or deviant behavior, including shoplifting, vandalism, truancy, and fighting, which they found correlated .41 with the number of times an adolescent male has been drunk in the past year. The relation is similar for high school and college males. Jessor has suggested (see Donovan & Jessor, 1985) that these behaviors cluster together because they represent unconventional behavior, or behavior that is undesirable by the norms of conventional society. To a certain extent, both frequency of being drunk and the composite of deviant behaviors are negatively correlated with two indices of conventional behavior, namely church attendance and school performance. These two have frequently been cited as protective factors.

A program of research by Pihl and colleagues (e.g., Pihl, Peterson, & Finn, 1990) documents support for the claim that specific personality, cognitive, and behavioral characteristics are more commonly found among sons of male alcoholics than among other boys, before they have actually begun to drink alcohol. The boys are more likely than matched controls to be described as impulsive, hyperactive, conduct disordered, and aggressive. Similarly, Brook et al. (1992) found that the best predictor of alcohol abuse in 16–21 year old Whites was early aggression. The heavy drinking male adolescent with a hyperactive, aggressive, or impulsive temperament from youth, and alcoholism in the family is sometimes referred to as having Type II alcoholism. Many researchers believe that a large part of the predisposition, to conduct disorder or to alcohol abuse, is genetic.

Type I alcoholics, on the other hand, are male or female and their alcoholism is thought to be caused by an interaction of many factors, personality and environmental. The stimulating properties of alcohol may be more rewarding for them, especially if they fit the profile of sensation seekers or are anxious or depressed. The stimulation may come not only from the alcohol, but also from the challenge to authority and the risky peer group activities. Risk-taking behaviors are more common in adolescence than at other times in the lifespan, and serve specific psychosocial functions: belonging to a peer group, having fun, showing independence from parents and other adults, and coping with feelings of inadequacy and stress. As Tonkin, Cox, Blackman, and Sheps (1990) point out, for most adolescents, drinking serves many of these functions with a minimum of risk. However, for a subgroup of adolescents, drinking is part of a cluster of high-risk behaviors that include driving while intoxicated, precocious sexuality, and delinquency. These adolescents appear to have higher levels of unconventionality as measured by Zuckerman's (1979) Sensation-Seeking Scale. Sensation seeking has four components: risk taking, disinhibition, adventure seeking, and susceptibility to boredom. One subscale in particular is related to drinking to intoxication, namely the Disinhibition subscale which measures rebellion against strict codes of acceptable behavior. People who seek excitement but are inhibited may use alcohol to disinhibit themselves enough to engage in the exciting activities. Tonkin et al. (1990) believe that this group sets itself apart from most adolescents in terms of the cluster of high risk-taking behaviors they engage in at an earlier age.

Social Environment

Certain social environments are highly conducive to heavy weekend drinking during adolescence. Over 70% of White students and over 50% of Indian students said that they drink to have fun at a party (Binion, Miller, Beauvais, & Oetting, 1988). Likewise, many said it made them feel part of a group of friends. The other common reasons concerned the new, exciting sensations and the altered drugged state they experienced. The latter reasons are relevant to our previous discussion of characteristics of the drinker, yet it is clear that the experiences are more positive in the company of others than alone. Indian students also said they drank to feel free and independent and out of boredom.

The influence of peers is also apparent among those living with peers in a college dormitory. Follow-up studies of seniors after they leave high school show that drinking reaches a peak in the 18–20 year period, and is particularly high among those living in a dormitory or group setting. This group showed the lowest levels of drinking when in high school and will again drop to low levels when they leave to adopt adult roles in marriage and employment (Bachman et al., 1997). However, the combination of closeness to peers and availability of inexpensive beer contributes to heavy drinking during the single years.

Others regard peer norms and more specifically the activities and values of close friends to be influential in adolescence (e.g., Holden et al., 1990). This is particularly true in the middle teen years, but conformity is lower in late adolescence when drinking tends to be high (Gavin & Furman, 1989). Even in the mid-teen years, groups are not a source of support the way friends are, but a source of reputation, status, conformity, and conflict — a very ambivalent mix. Rather than being a one-directional influence on the naive adolescent, peers are likely to be selected as friends to the extent that they share similar interests. The influence is bi-directional. The suggestion that peer influence is sometimes less than expected was highlighted in the reports of adolescents who had driven while intoxicated (Vegega & Klitzner, 1989). Peer pressure to drink or to drive was minimal according to the respondents; only 15% said they were pressured to drink either through encouragement or a drinking competition, and only 13% were pressured to drive by someone who needed to get home.

Family and school stress are sometimes considered as causes or correlates of alcohol abuse among adolescents (Clayton, 1992). Forney et al. (1984) found no evidence for drinkers coming from homes with marital breakdown. Others have found that Black teens may have more of a drinking problem as a result of their mother being alone, not as a result of their father being absent (Rhodes & Jason, 1990). This implies that the family stress arising from lack of economic and social support for the mother leads children to cope with peer and school demands on their own. When they are unable to cope, they use alcohol to relieve negative emotional states (anxiety and depression). School stress, likewise, may lead to alienation from an important socializing institution, from important authority figures, and from socially engaged peers. For these reasons, preventive programs help to strengthen adolescents' involvement in the school and family.

Considering a number of the individual, family, and peer influences on adolescent drinking, Brook et al. (1992) examined the predictive value of each in a longitudinal design. Their sample was mostly White Americans from upstate New York between the ages of 16 and 21. A certain cluster of variables predicted moderate use (compared to light or non-use); a different cluster predicted abuse compared to moderate use. Moderate use was best predicted by peer and social orientation variables, such as alcohol use by friends, time spent with friends, peer deviance, tolerance of deviance, and sensation seeking. Abuse was best predicted by childhood aggression, which included ratings of anger, aggression to siblings, noncompliance, temper, and nonconforming behavior, and by acting out as an adolescent (rebellious, impulsive, deviant), low attachment to and identification with mother, and illegal drug use among peers. Parent drug and alcohol use and low-achieving friends were not predictive.

PREVENTION PROGRAMS

Although treatment programs have been conducted with adolescent alcoholics, the preferred intervention is a preventive program (Moser, 1980). Prevention is aimed at reducing alcohol abuse, reducing the chances that the adolescent will become dependent on alcohol, and reducing the social and physical consequences of abuse. They take two broad approaches: one is through social control of exposure to alcohol and the other is by instilling personal control through school- or family-based education. The social control strategy is a public health approach to prevention that aims to control the exposure to and availability of alcohol at the community level. The education strategy is directed at adolescents with the goal of changing their individual response to alcohol.

Controlling Availability of Alcohol

On the basis of survey data, governments may seek to control the availability of and exposure to alcohol as a way of reducing intake among adolescents (e.g., Bucholz & Robins, 1989; Howard, Ganikos, & Taylor, 1990; Kortteinen, 1989; Moser, 1980). Governments have the authority to control the distribution, price, and advertisement of alcoholic beverages by enacting and enforcing laws. Whether or not the motives are to control access to alcohol or to profit financially, controls serve to minimize the harmful effects of drinking without infringing on the rights of those who use alcohol moderately and responsibly. For example, there is a partial government monopoly on the sale of alcohol in Canada through provincial government distribution stores. Only low concentration beverages, such as wine and beer, may be sold at licensed private outlets. Other countries around the world vary in the level of state control over access to alcohol, with some having control over production as well as distribution and others over licensing distribution only (Kortteinen, 1989).

To minimize the use of alcohol by adolescents, governments may restrict sales to people 21 years and over. If caught selling to under-age adolescents, the establishment may be fined or eventually have their licence revoked as a penalty. Investigative journalists periodically demonstrate with video evidence that this law is not always enforced. In addition, adults cannot be prevented from buying alcohol and giving it to the under-age person. The concept of control is more concerned with increasing the difficulty in acquiring large amounts of alcohol. As we saw previously, alcohol consumption reaches its peak around 18 to 20 years, the age when it is or was legal to buy. Consequently, most states and some provinces have raised the legal age from 18 to 21 years.

Another effective control strategy is to set prices and/or taxes beyond the means of youngsters and above the cost of essential food. This has been a very effective way of reducing alcohol consumption and the prevalence of alcoholism in the general population, while at the same time raising revenues for health or other services (Moser, 1980). However, where there are no minimum prices set, the desire for profit may motivate commercial establishments to lower their prices in an effort to capture the market.

Another strategy is to set the hours of sale and locations in the community where alcohol may be sold. Adolescent drinking usually takes place on weekends and in the evenings. If stores selling alcohol are required to close in the early evening on the weekend, then intoxicated adolescents will not be able to replenish their supplies mid-evening. In some communities, alcohol outlets are not allowed to be located near hospitals or schools. Certain Indian reservations have banned the sale of alcohol altogether, with mixed results.

Other governments focus their efforts at control less on the availability of alcohol and more on the negative consequences of intoxication. In this case, they will have laws that penalize vehicle drivers who are intoxicated. In most places, 0.10% is the level of blood alcohol concentration (BAC) at which a person is considered legally intoxicated, meaning 100 mg of alcohol per 100 ml of blood. However, rational thought and muscular control are impaired at lower levels of 0.05% (Davidson, 1985). Consequently, one option for controlling injuries caused by drinking and driving is to lower the legal limit to 0.08% as some places have already done.

Finally, a government can control advertising of alcohol. It can restrict advertising to places where adolescents will not be exposed to it and also minimize the image appeal. This form of control generally receives broad support from the adult population but meets with resistance from alcohol producers. The on-again–off-again controls on tobacco advertising in Canada demonstrate the difficulty in resolving a conflict between freedom of speech and freedom from addiction or intoxication. Advertising alcohol on billboards exposes children and adolescents repeatedly to the products. Advertising alcohol in older youth and adult magazines such as sports, music and fashion magazines exposes adolescents to these products. In addition, image-oriented advertising is particularly persuasive for those who identify themselves in terms of their external appearance. At an age when emblematic indicators of identity are important, many adolescents are persuaded by image-oriented advertising of young good-looking people drinking and having fun with the opposite sex (Covell, Dion, & Dion, 1994). Although no psychologist would claim that simply observing such an ad causes adolescents to drink excessively, the environmental exposure interacts with predispositions of the adolescents (such as conformity and identity confusion) to influence drinking behavior.

School- and Family-Based Education

Preventive programs in schools and with families have been credited with the decline in alcohol use among high school seniors since the mid-1980s (Holden et al., 1990). However, a review of these programs yields mixed results. Thus, the question still remains as to which types of interventions are more successful than others. Early preventive programs in the schools try to reach students in grades 7 and 8 with information about the harmful effects of alcohol (e.g., the Personal and Social Development curriculum in Quebec schools). Programs with older high school students target those at high risk or those with alcohol problems who have a record of driving while intoxicated.

Bry and Greene (1990) reviewed the school- and family-based programs. They aim to reduce problem drinking as well as the negative consequences that follow by changing circumstances that put students at risk for heavy drinking. The Life Skills Training program teaches students how to have more personal control over their drinking, for example, by learning how to resist peer pressure to drink (Holden et al., 1990). One such program, called Friendly PEERsuasion, gives girls 14 bi-weekly sessions in how to be more assertive with their persuasive boyfriends. The girls are trained to be trainers of other girls, thereby giving them status as responsible "teachers" of younger girls. Through discussion and practice, they role play situations where boys try to persuade them to drink — and their replies are not always a polite "no thank you" (Chaiken, 1990). Another program, called The Early Secondary Intervention Program, provides two years of weekly meetings led by school personnel to improve self-discipline and school performance. Absenteeism, tardiness, incomplete homework, poor grades, and trouble with teachers are directly addressed with each student after receiving feedback from teachers. Rewards are given to students who show improvement. After two years, the evaluation indicated improvement in school behavior, more paid employment, and fewer arrests among students in the program compared to those not in the program (Bry & Greene, 1990). However, alcohol use did not differ for intervention compared to controls. Consequently, the program has expanded to include family involvement.

Families were involved with a structured program called Problem-Solving Communication. This stems from the research showing that adolescent problem drinking is often a reaction to conflict with parents and a defiant bid for independence from parental constraints. Parents are counselled in how to negotiate rules with their adolescent and how to reinforce compliance with rules. It also teaches parents how to use positive communication (e.g., Bowman & Howard, 1985). At the same time, parents are helped in solving their own family and personal problems. The stated goal is to motivate and train parents to teach their adolescent how to handle school and community demands. The program has been successful in including hard-to-reach families and their adolescents by holding meetings at school and at home where parents and their teens feel in control. Booster sessions are given at follow-up to those who have shown little improvement, and they are effective with some adolescents. By including the families in this type of counselling, the programs have shown greater success in both reducing drinking and improving school performance (Bry & Greene, 1990). Family-based intervention is also considered the best way to help Black adolescents deal with drinking problems and the school and communitiy problems associated with drinking (Oyemade & Washington, 1990).

In summary, prevention programs attempt to impose social control over the exposure to or availability of alcohol, or to teach personal control through skills training and peer and family supports. Raising the price of alcohol and the age of legal drinking are successful public health measures that limit the availability of alcohol. Media control over exposure to alcohol has not been implemented in a general way, but initial success with tobacco advertising indicates that this route may also be successful with alcohol. School-based skills training is based on the recognition that peer pressure is an exacerbating factor in adolescent problem drinking, even though it may not be a cause in itself. Often the most successful programs are those led by peers because they reinforce the idea that controlled drinking is the norm among responsible adolescents. Research on the risk factors associated with problem drinking identified other unconventional behaviors such as school delinquency and the selection of friends with similar problems. In recognition of this risk factor, many school-based interventions have successfully improved the school performance and discipline of students, so that they do not become involved in activities that lead to alcohol abuse and its related problems. Finally, in recognition of the family factors that lead to drinking, such as alienation from parents, family-based programs deal with parent-adolescent conflict resolution.

Sensible and responsible drinking is only one of the goals of prevention programs. For the most part, school and health workers see adolescent drinking as a marker for other personal and social problems. By working in family and school settings, they are able to minimize the negative consequences of drinking, as well as the risk factors for drinking.

REFERENCES

Aronson, M. D. (1987). Definition of alcoholism. In H.N. Barnes, M.D. Aronson & T.L. Delbanco (Eds.), *Alcoholism: A guide for the primary care physician* (pp. 9–15). New York: Springler-Verlaz.

Bachman, J. G., Wadsworth, K. N., O'Malley, P. M., Johnston, L. D., & Schulenber, J. E. (1997). *Smoking, drinking and drug use in young adulthood*. Hillsdale, NJ: Erlbaum.

Binion, A., Miller, C. D., Beauvais, F., & Oetting, E. R. (1988). Rationales for the use of alcohol, marijuana, and other drugs by eighth grade Native American and Anglo youth. *The International Journal of the Addictions, 23*, 47–64.

Bowman, P. J., & Howard, C. (1985). Race-related socialization, motivation, and academic achievement: A study of Black youth in three-generation families. *Journal of the American Academy of Child Psychiatry, 24*, 134–141.

Brook, J. S., Cohen, P., Whiteman, M., & Gordon, A. S. (1992). Psychosocial risk factors in the transition from moderate to heavy use or abuse of drugs. In M. Glantz & R. Pickens (Eds.), *Vulnerabilty to drug abuse* (pp. 359–388). Washington DC: American Psychological Association.

Brunswick, A. F., Messeri, P. A., & Aidala, A. A. (1990). Changing drug patterns and treatment behavior: A longitudinal study of urban Black youth. In R. R. Watson (Ed.), *Drug and alcohol abuse prevention* (pp. 263–311). Clifton, NJ: Humana Press.

Bry, B. H., & Greene, D. M. (1990). Empirical bases for integrating school- and family-based interventions against early adolescent substance abuse. In R. J. McMahon & R. D. Peters (Eds.), *Behavior disorders of adolescence: Research, intervention, and policy in clinical and school settings* (pp. 81–97). New York: Plenum Press.

Bucholz, K. K., & Robins, L. N. (1989). Sociological research on alcohol use, problems, and policy. *Annual Review of Sociology, 15*, 163–186.

Caspi, A., Elder, G. H., & Bem, D. J. (1987). Moving against the world: Life-course patterns of explosive children. *Developmental Psychology, 23*, 308–313.

Chaiken, M. R. (1990). Evaluation of Girls Clubs of America's friendly PEERsuasion program. In R. R. Watson (Ed.), *Drug and alcohol abuse prevention* (pp. 95–132). Clifton, NJ: Humana Press.

Clayton, R. R. (1992). Transitions in drug use: Risk and protective factors. In M. Glantz & R. Pickens (Eds.), *Vulnerability to drug abuse* (pp. 15–51). Washington DC: APA.

Compton, W. M., Helzer, J. E., Hwu, H. G., Yeh, E. K., McEvoy, L., Tipp, J. E., & Spitznagel, E. L. (1991). New methods in cross-cultural psychiatry: Psychiatric illness in Taiwan and the United States. *American Journal of Psychiatry, 148*, 1697–1704.

Covell, K., Dion, K. L., & Dion, K. K. (1994). Gender differences in evaluations of tobacco and alcohol advertisements. *Canadian Journal of Behavioral Science, 26*, 404–420.

Davidson, R. S. (1985). Behavioral medicine and alcoholism. In N. Schneiderman & J. T. Tapp (Eds.), *Behavioral medicine: The biopsychosocial approach* (pp. 379–404). Hillsdale, NJ: Erlbaum.

Dewit, D. J., Silverman, G., Goodstadt, M., & Stoduto, G. (1995). The construction of risk and protective factor indices for adolescent alcohol and other drug use. *Journal of Drug Issues, 25*, 837–863.

Donovan, J. E., & Jessor, R. (1985). Structure of problem behavior in adolescence and young adulthood. *Journal of Consulting and Clinical Psychology, 53*, 890–904.

Eliany, M. (1992). Alcohol and drug consumption among Canadian youth. *Statistics Canada: Canadian Social Trends*, Autumn, 10–13.

Fillmore, K. M. (1988). *Alcohol use across the life course: A critical review of 70 years of international longitudinal research.* Toronto: Addiction Research Foundation.

Fillmore, K. M., Hartka, E., Johnstone, B. M. Leino, E. V., Motoyoshi, M., & Temple, M. T. (1991). A meta-analysis of life course variation in drinking. *British Journal of Addiction, 86*, 1221–1268.

Forney, M. A., Forney, P. D., Davis, H., Van Hoose, J., Cafferty, T., & Allen, H. (1984). A discriminant analysis of adolescent problem drinking. *Journal of Drug Education, 14*, 347–355.

Freeman, E. M. (1990). Social competence as a framework for addressing ethnicity and teenage alcohol problems. In A. R. Stiffman & L. E. Davis (Eds.), *Ethnic issues in adolescent mental health* (pp. 247–266). Newbury Park, CA: Sage.

Gavin, L. A., & Furman, W. (1989). Age differences in adolescents' perceptions of their peer groups. *Developmental Psychology, 25*, 827–834.

Giovino, G. A., Henningfield, J. E., Tomar, S. L., Escobedo, L. G., & Slade, J. (1995). Epidemiology of tobacco use and dependence. *Epidemiologic Reviews, 17*, 48–65.

Glantz, M. D., & Pickens, R. W. (1992). Vulnerability to drug abuse: introduction and overview. In M. Glantz & R. Pickens (Eds.), *Vulnerability to drug abuse* (pp. 1–14). Washington DC: APA.

Gower, D. (1990). Under the influence. *Statistics Canada: Perspectives*, Autumn, 30–41.

Helzer, J. E., Canino, G. J., Yeh, E. K., Bland, R. C., Lee, C. K., Howes, H. G., & Newman, S. (1990). Alcoholism — North America and Asia. *Archives of General Psychiatry, 47*, 313–319.

Holden, G. W., Moncher, M. S., & Schinke, S. P. (1990). Substance abuse. In A. S. Bellack, M. Hersen, & A. E. Kazdin (Eds.), *International handbook of behavior modification and therapy* (2nd ed., pp. 869–880). New York: Plenum Press.

Howard, J., Ganikos, M. L., & Taylor, J. A. (1990). Alcohol prevention research: Confronting the challenge. In R. R Watson (Ed.), *Drug and alcohol abuse prevention* (pp. 1–18). Clifton, NJ: Humana Press.

Hurley, J., & Horowitz, J. (1990). *Alcohol and health.* New York: Hemisphere Publications.

Johnston, L. D. (1985). The etiology and prevention of substance use: What can we learn from recent historical changes? In C. L. Jones & R. J. Battjes (Eds.), *Etiology of drug abuse: Implications for prevention* (pp. 155–177). Washington DC: US Government.

Kandel, D. B., & Logan, J. A. (1984). Patterns of drug use from adolescence to young adulthood: I. Periods of risk for initiation, continued use, and discontinuation. *American Journal of Public Health, 74*, 660–666.

Kandel, D. B., & Yamaguchi, K. (1985). Developmental patterns of the use of legal, illegal, and medically prescribed psychotropic drugs from adolescence to young adulthood. In C. L Jones & R. J. Battjes (Eds.), *Etiology of drug abuse: Implications for prevention* (pp. 193–235). Washington DC: US Government.

Kortteinen, T. (1989). State monopoly systems and alcohol prevention in developing countries: Report on a collaborative international study. *British Journal of Addictions, 84*, 413–425.

Millar, W. (1991). A trend to a healthier life. *Health Reports, 3*, 363–370.

Miller, W. R., Heather, N., & Hall, W. (1991). Calculating standard drink units: International comparisons. *British Journal of Addiction, 86*, 43–47.

Moncher, M. S., Holden, G. W., & Trimble, J. E. (1990). Substance abuse among Native-American youth. *Journal of Consulting and Clinical Psychology, 58*, 408–415.

Moser, J. (1980). *Prevention of alcohol-related problems: An international review of preventive measures, policies and programmes.* Geneva: WHO.

Murdoch, D., Phil, R. O., & Ross, D. (1990). Alcohol and crimes of violence: Present issues. *The International Journal of the Addictions, 25*, 1065–1081.

Murray, D. M., Perry, C. L., O'Connell, C., & Schmid, L. (1987). Seventh-grade cigarette, alcohol and marijuana use: Distribution in a north central U.S. metropolitan population. *The International Journal of the Addictions, 22*, 357–376.

Newcomb, M. D., Maddahian, E., & Bentler, P. M. (1986). Risk factors for drug use among adolescents: Concurrent and longitudinal analyses. *American Journal of Public Health, 76*, 525–531.

Oetting, E. R., & Beauvais, F. (1990). Adolescent drug use: Findings of national and local surveys. *Journal of Consulting and Clinical Psychology, 58*, 385–394.

O'Nell, T. D., & Mitchell, C. M. (1996). Alcohol use among American Indian adolescents: The role of culture in pathological drinking. *Social Science & Medicine, 42*, 565–578.

Oyemade, U. J., & Washington, V. (1990). The role of family factors in the primary prevention of substance use and abuse among high risk black youth. In A. R. Stiffman & L. E. Davis (Eds.), *Ethnic issues in adolescent mental health* (pp. 267–284). Newbury Park, CA: Sage.

Pihl, R. O., & Peterson, J. B. (1992). Etiology. *Annual review of addictions research and treatment*, 153–175.

Pihl, R. O., Peterson, J., & Finn, P. (1990). Inherited predisposition to alcoholism: Characteristics of sons of male alcoholics. *Journal of Abnormal Psychology, 99*, 291–301.

Rhodes, J. E., & Jason, L. A. (1990). A social stress model of substance abuse. *Journal of Consulting and Clinical Psychology, 58*, 395–401.

Robins, L. N., & Przybeck, T. R. (1985). Age of onset of drug use as a factor in drug and other disorders. In C. L. Jones & R. J. Battjes (Eds.), *Etiology of drug abuse: Implications for prevention* (pp. 178–192). Washington DC: US Government.

Tonkin, R. S., Cox, D. N., Blackman, A. R., & Sheps, S. (1990). Risk-taking behavior in adolescence. In R. J. McMahon, & R. D. Peters (Eds.), *Behavior disorders of adolescence: Research, intervention, and policy in clinical and school settings* (pp. 27–37). New York: Plenum Press.

Vegega, M. E., & Klitzner, M. D. (1989). Drinking and driving among youth: A study of situational risk factors. *Health Education Quarterly, 16*, 373–388.

Walker, R. D., Lambert, M. D., Walker, P. S., & Kivlahan, D. R. (1993). Treatment implications of comorbid psychopathology in American Indians and Alaska Natives. *Culture, Medicine and Psychiatry, 16*, 555–572.

Wilsnack, R. W., & Wilsnack, S. C. (1978). Sex roles and drinking among adolescent girls. *Journal of Studies on Alcohol, 39*, 1855–1874.

Zuckerman, M. (1979). *Sensation seeking: Beyond the optimal level of arousal.* Hillsdale, NJ: Erlbaum.

8

SELF-REGULATION AS A KEY TO SUCCESS IN LIFE[*]

Roy F. Baumeister, Karen P. Leith, Mark Muraven, and Ellen Bratslavsky

Case Western Reserve University

Self-regulation is one of the most important traits in the human psyche. Some other species have limited capacities for self-regulation, but this ability is far more developed and powerful in human beings. It is largely responsible for the immense diversity of human behavior.

Self-regulation (or the very similar term *self-control*) can be defined as the ability to alter one's own behavior, including one's thoughts, feelings, actions, and other responses. To put self-regulation in context, it is useful to step back for a moment and realize how far psychology came working with stimulus–response models. A great deal of human and animal behavior is, indeed, simply responses to stimuli. What stimulus–response theories miss, however, is the possibility of altering one's response. Human beings have a remarkably powerful ability to prevent themselves from responding to a stimulus in the normal or natural way.

To illustrate, consider an easy example of stimulus and response: A tempting piece of steak is set in front of a hungry dog. The dog's response is easy to predict: He will devour the steak. But put the same steak in front of an equally hungry human being, and the outcome is harder to predict. The impulse to eat the steak may be there, but the person may easily override that response, for a variety of other reasons: being on a diet, being a vegetarian, worries about cholesterol, and the like. Nothing would be more natural than eating a delicious food when one is hungry, but many people frequently override that response.

BENEFITS OF SELF-REGULATION

The title of this chapter asserts that self-regulation is a key to success in life. Success in life was defined by Dolores Pushkar at the 1996 Concordia Conference on Competence

[*] Address correspondence to R. Baumeister, Dept. of Psychology, 10900 Euclid Ave., Cleveland, Ohio 44106-7123, rfb2@po.cwru.edu. Preparation of this chapter, and some of the research described in it, were supported and facilitated by grant #MH 51482 from the National Institutes of Health.

Improving Competence across the Lifespan, edited by Pushkar *et al.*
Plenum Press, New York, 1998.

Through the Life Span as a matter of being able to live with oneself and to live with others. If self-regulation is indeed a key to success in life, then it should improve people's ability to live with themselves and with others. The research reviewed in subsequent sections of this chapter will contribute to making this point, but it is worth adding some general observations here. These are general conclusions based on extensive literature reviews concerning failures of self-regulation (Baumeister, Heatherton, & Tice, 1994).

With regard to improving how well one can live with oneself, self-regulation has several benefits. Self-regulation encompasses affect regulation, that is, the control over one's emotional states and moods, and it is easy to appreciate that success at controlling emotions will enable one to feel better and suffer less on a daily basis. Self-regulation also includes control over one's mental processes, such as the ability to concentrate and to persist on tasks. After all, the natural "response" to the "stimulus" of numbers on a page is hardly to do arithmetic exercises, but if a child is to grow up as a successful and competent member of modern society, he or she is going to have to override the natural response and do math homework at some point.

Controlling impulses is another important sphere of self-regulation, and it too contributes to success in life. The most obvious example of the problems of self-regulation failure in this sphere is addiction to drugs or alcohol. Self-regulation enables people to resist a variety of temptations.

Self-regulation also involves setting and reaching goals. To succeed in life, people must manage themselves effectively, which involves setting appropriate goals and then making themselves carry out the steps to achieve them. Often this involves persisting in the face of failures or setbacks. Self-regulation is crucial for enabling people to do this.

Self-defeating behaviors constitute some of the most significant ways in which people fail to live with themselves (Baumeister & Scher, 1988). Self-defeating behavior involves whatever people do that thwarts their strivings or brings suffering, misfortune, and failure on themselves. A great deal of self-regulation consists of preventing self-defeating behaviors — in fact, the general patterns of self-defeating behavior offer multiple parallels to the general principles of self-regulation failure (Baumeister, 1997a).

Turning from the issue of living with oneself to living with others, it is readily apparent that self-regulation is again helpful. Controlling emotions is often just as important and valuable for helping one get along with others as it is for securing one's own affective serenity. Likewise, self-regulation enables people to keep their promises and fulfill their obligations when they might not feel like doing so or might be tempted to act otherwise.

Addiction was mentioned as one sphere in which poor self-regulation contributes to misfortune and suffering for the self. Addiction also has an interpersonal aspect, of course. It is often the families of addicts and alcoholics who suffer immensely, and addictive processes are quite destructive of family life and friendship. More generally, self-regulation can reduce infidelity, betrayal, and other behaviors that involve yielding to temptations in ways that harm close relationships.

Violence is probably the most destructive interpersonal behavior. Violent behavior often follows from failures of self-regulation. Gottfredson and Hirschi (1990) proposed a general theory of crime that revolved around poor self-control. Baumeister (1997b) noted there are so many causes of violence and aggression, it is surprising that there is not more violence than there is — and the reason is that most violent impulses are restrained by inner inhibitions. Thus, self-regulation is important for preventing violence.

Although this brief survey is sufficient to suggest the range of benefits of self-regulation, it is worth citing one important study that provided longitudinal evidence of such benefits. Mischel, Shoda, and Peake (1988; see also Shoda, Mischel, & Peake, 1990) measured children's ability to delay gratification (i.e., to refuse immediate rewards for the sake of obtaining larger, but delayed rewards) when they were 4 or 5 years old. Delay of gratification is an important and basic form of self-regulation. Over a decade later, the researchers found that the children who had showed the best capacity for self-regulation went on to be the most successful in young adult life. They were superior to others in terms of school performance and college readiness, in terms of social competence and getting along with others, and in terms of personal strengths such as being able to cope with frustration and stress effectively. These results suggest that self-regulation is a central aspect of personality that is stable across many important developmental changes and consistently yields positive outcomes that benefit both the individual and the social network.

Thus, it is safe to say that self-regulation is centrally involved in many activities that hold the possibility to make people happy and successful or miserable and unsuccessful in life. In the next sections, we shall cover what our own research efforts have learned about self-regulation.

SETTING AND REACHING GOALS

One set of definitions for success in life focuses on how well people can reach their goals (e.g., Gollwitzer & Bargh, 1996). Yet reaching goals is obviously not a simple matter. Two separate processes must be understood, and both can involve self-regulation. The first is a matter of setting appropriate goals. The second is a matter of pursuing them effectively and persistently so as to achieve success.

The implications for success in life are important. Two people with identical levels of academic ability may end up performing quite differently and having very different grade point averages over a couple years of college, if one is better at this sort of self-management. By choosing courses appropriate to one's level of ability and by budgeting one's time and effort properly, one can gain the maximum return for one's ability. In contrast, an equally intelligent person who selects courses that are alternately far too hard or too easy will end up with a poorer education and lower performance. Thus, by selecting appropriate contingencies and setting proper goals and obligations for oneself, one can maximize one's successes.

Although many laboratory experiments set explicit goals for their participants, in everyday life people are often called upon to set their own goals. How one sets one's goals can have considerable impact on whether one reaches them or not. After all, some goals are presumably so simple that anyone can easily reach them and success is almost guaranteed, whereas others may be practically impossible and hence failure is ensured.

If life were simply a matter of reaching one's goals, then the wisest advice would be to set extremely low goals, because these have the highest probability of success. Yet obviously people do not do this. The reason is that goals have various rewards and subjective values associated with them, and easy goals tend to have low values. For example, one's chances of securing a mate may be highest if one fixes one's romantic aspirations on someone so undesirable that one will have no rivals and the person will presumably be desperately grateful for any attention one shows. One has thus a high chance of success — but the success may be worth relatively little, because the mate is, by definition, quite undesirable. A similar logic applies to career aspirations and other spheres of endeavor.

Setting goals thus requires recognizing a tradeoff between the value of success and the likelihood of success. The most valuable goals are usually those with the lowest likelihood of success. Somewhere along that continuum the person must find the optimal balance. Ideally this will be at the point at which one's own abilities and other qualities are sufficient to keep the likelihood of success high while the value is also still high. In simple terms, one ideally wants the best goal that one has a good chance of reaching.

Self-knowledge thus emerges as a key factor in setting goals. To set goals effectively, one needs to know how much one can accomplish. In pursuing a mate, for example, a person would ideally have a good understanding of his or her own attractiveness and other factors that can contribute to one's romantic appeal, so as to be able to appraise one's chances that a particular mate will regard one as suitable.

There is, however, a further complication to the matter of setting goals on the continuum that ranges from easy but worthless up to wonderful but impossible. Given the difficulty of predicting exactly how well one can do, it is to be expected that people will periodically overshoot or undershoot; that is will set goals that are above or below the optimal point. The complication is that there is an asymmetry to the consequences of those two types of error. Overshooting can often be significantly worse than undershooting.

Undershooting means setting a goal that is somewhat below the best one could possibly achieve. Such a goal means that one is very likely to achieve success, but the success will not have the highest value one could achieve. In contrast, overshooting means setting too high a goal, which will normally result in failure. Thus, undershooting brings a slightly diminished success, whereas overshooting brings failure. Failure is often considerably worse than a slightly diminished success.

For this reason, perhaps, popular wisdom offers an assortment of sayings that recommend selecting a goal that is somewhat lower than the best possible (so as to "leave a margin for error," for example). The optimal strategy, in other words, would be to be slightly underconfident. Yet this strategy seemingly conflicts with the body of research findings suggesting that in terms of self-knowledge and self-prediction, people tend toward broad patterns of overconfidence (Taylor & Brown, 1988; Vallone, Griffin, Lin, & Ross, 1990). Given the broad tendencies toward positive illusions and inflated self-esteem, people should be prone to make the more dangerous kind of mistake, namely setting goals that are too high.

A series of studies examined this dilemma of overconfident goal-setting (Baumeister, Heatherton, & Tice, 1993). To include the role of positive illusions, we measured self-esteem. We also included one condition with an ego threat, because previous findings had shown that inflated self-appraisals and predictions may be especially common in people with high self-esteem who have received an ego threat (McFarlin & Blascovich, 1981). We had people perform a task (a video game) through a long learning phase, during which they recorded their scores (thereby facilitating self-knowledge about their capabilities). Then we had them perform trials with money on the line. In one study we had them select goals for themselves, such that higher goals carried greater financial rewards. In two other studies, we selected a target goal that was at the 67th percentile of the subject's own scores, and we allowed the subject to bet, at triple or nothing, any part of $3 we paid him or her on the subject's chances of surpassing that criterion on the final trial. In both procedures, the subject had to make a fairly accurate prediction of how well he or she would perform and set the contingency accordingly. The situation included the asymmetry of consequences of overshooting versus undershooting mentioned earlier, insofar as if the person failed to reach the criterion, all money (at least all that was bet) was lost.

The results of these studies shed interesting light on how people set goals and contingencies for themselves. The bottom line in all studies was how much money the subject managed to earn in the experiment, because that was the outcome measure of self-regulation. In the betting studies, the subject could do well by making a low bet and keeping most or all of the $3, and the subject could do very well by making a high bet and then performing well, in which case he or she could gain up to $9. The subject only fared badly by making a large bet and then not performing well, which would entail losing the stake.

In the condition where no ego threat was involved, people with high self-esteem did well and ended up with an impressive amount of money — indeed, significantly more than people with low self-esteem earned. This finding suggests that when times are good, people with high self-esteem manage themselves effectively, in the sense that they set appropriate goals and then perform up to their own expectations. These results fit Campbell's (1990; see also Campbell & Lavallee, 1993) findings that people with high self-esteem have superior self-knowledge and process information bearing on the self better than people with low self-esteem. In our study, apparently, people with high self-esteem learned more quickly and accurately what they were capable of doing and were able to set their own goal contingencies accordingly.

This effective self-management pattern was thoroughly disrupted, however, when people with high self-esteem received an ego threat. In two studies, the ego threat consisted of the experimenter intimating that the subject might not "have what it takes to perform well under pressure," and in a third it consisted of randomly assigned failure feedback on a creativity test (that was ostensibly unrelated to the video game task). People with high self-esteem responded to such threats with extremely positive, self-aggrandizing assertions, including setting very high goals for themselves or making maximum bets on their own performance. These highly self-confident responses were often unwise, in that they exceeded what the subject was likely to achieve, and so these individuals tended to lose all their money. On average, they left the experiment with the lowest earnings among all the cells, and indeed their average take-home pay was significantly lower than what people with low self-esteem earned in the same condition.

Thus, people with high self-esteem showed the best and the worst self-management in this study, as measured in terms of how much money they managed to earn by setting reward contingencies for their own performance. When things had been going well, they set appropriate goals and performed well, and they earned an average of five and a half dollars. When they received an ego threat, however, they set overconfident goals and failed to live up to them, and their take-home earnings averaged a paltry 93 cents (out of $9 maximum).

There is thus nothing inherently wrong with having a favorable opinion of oneself. People with high self-esteem did manage themselves quite well in the condition where there was no ego threat. Taylor and Brown (1988) have proposed that positive illusions, in the sense of holding highly favorable and possibly inflated views of oneself, are generally adaptive, and our results supported that view — at least in the prior success condition. On the other hand, an ego threat seemed to undermine the self-managing efficacy of people with high self-esteem. Apparently their response to an esteem threat is immediately to assert their superior capability. These responses are unwise, however, and the overly confident commitments these individuals make tend to backfire in a costly fashion.

These results also fit a broader pattern suggesting that threatened egotism is a particularly dangerous condition. In this study, people who thought highly of themselves but encountered an external threat to their favorable self-image made foolish, risky choices that ended up costing them money. Other work has associated threatened egotism with violent, aggressive behaviors (Baumeister, Smart, & Boden, 1996) and with various self-defeating responses (Baumeister, 1997a). An inflated self-opinion plus an external esteem threat may be a general recipe for destructive, problematic responses.

PROCRASTINATION

Doing one's work on time and fulfilling other obligations in a timely fashion would seem to be an integral part of healthy, proper adult functioning. Yet the majority of people report that they procrastinate on some things, and a substantial minority of people report that their procrastination habits are serious enough to cause personal, financial, or occupational problems for them (Ferrari, Johnson, & McCown, 1995).

Procrastination is often criticized, especially by people who do not see themselves as guilty of it, as a form of self-regulatory failure indicative of laziness, self-indulgence, and poor self-management. These critics contend that putting things off until the last minute will tend to lower the quality of work, because one has to perform in a rushed manner. They are also sometimes self-righteous about the stress and other problems that procrastinators must endure when the deadline is looming (see Boice, 1989, 1996).

On the other hand, procrastination does have its apologists. Some point out quite plausibly that one can put the same amount of time into a project, resulting in the same quality of work, regardless of whether one does so early or late in the deadline period. Some procrastinators even contend that procrastination may actually improve their performance: "I do my best work under pressure" is a common statement by such people (Ferrari, 1992; Ferrari et al., 1995; Lay, 1995), implying that they raise the pressure by putting things off until the last minute. Other apologists say that the last-minute stress suffered by procrastinators should be balanced against a carefree, casual enjoyment of life at other times, in contrast to the possibly compulsive, driven, pervasively stressed style of the nonprocrastinators who always get right to work on any task. Some researchers draw parallels to the Type A personality, which is marked by constant drive and ambition (and presumably by a lack of competitiveness) — and by a tendency to suffer from heart disease which the more casual and carefree Type B is less prone to experience.

To investigate these competing hypotheses, Tice and Baumeister (in press) conducted a series of longitudinal studies. They assigned students a term paper long in advance and measured their procrastination tendencies using standard self-report measures (Lay, 1986; McCown & Johnson, 1989, 1991). The scales proved valid: self-described procrastinators did indeed turn their papers in later than other students.

Of greater interest, however, were the effects on performance and on health. Performance was assessed in terms of grades on the term paper, the midterm, and the final examination. All this work was graded by instructors who were blind to the students' procrastination status. In two studies, procrastinators achieved consistently lower grades on all measures. The difference was about two-thirds of a letter grade (a high B versus a C plus). Thus, the view that procrastination helps people do their best work by creating pressure appears to be false, as is the view that procrastination is innocuous. Other studies have suggested that procrastinators are just as intelligent as nonprocrastinators. The lower grades in these studies must, therefore, be attributed to the deleterious effects of procrastination.

Although procrastination may be bad for performance, the first study suggested that it may benefit health. Students in that study recorded their health symptoms over a 4-week period early in the semester, and procrastinators emerged as healthier than nonprocrastinators. This finding supports the view that procrastinators do derive some benefits from putting things off. It casts procrastination as a possible tradeoff, in which health and quality of life are improved in exchange for lower performance.

The ambiguity in those findings came from the fact that the health data were collected early in the semester, when the deadline was still remote. It seemed necessary to measure health late in the semester, when the procrastinator is presumably struggling to complete all the tasks that have been postponed. The second study included such a measure (as well as recording visits to the student clinic). This study replicated the finding from the first study that procrastinators are healthier early in the semester. The late-semester data showed, however, that procrastinators are considerably sicker when the deadline looms. Indeed, adding the early and the late data together, procrastinators emerged as sicker overall.

Taken together, these results portray procrastination as a self-defeating behavior pattern that should be included among the problems resulting from poor self-regulation. The net long-term effects of procrastination include harm to one's health and harm to one's performance. The only benefit is the short-term advantage to one's health that arises early in the performance period (but is then outbalanced by the higher illness later on). The pattern of short-term gain but higher long-term cost is one that characterizes many forms of self-defeating behavior (Baumeister & Scher, 1988; Platt, 1973) as well as being one common pattern of self-control failure (Baumeister, 1997a; Baumeister et al., 1994; Mischel, 1974, 1996; Mischel, Shoda, & Rodriguez, 1989).

The link of procrastination to self-defeating behavior raises the broader issue of how self-regulation contributes to such problems. The next section will examine one important link.

EMOTIONAL DISTRESS, RISK-TAKING, AND SELF-DESTRUCTIVE ACTS

The terms *self-defeating* and *self-destructive* are used (usually synonymously) to characterize behavior patterns in which people bring misfortune, failure, and suffering to themselves or otherwise prevent themselves from reaching their positive goals (e.g., Baumeister, 1997a; Baumeister & Scher, 1988; Berglas & Baumeister, 1993). Such behaviors have fascinated psychologists for many decades, in part because they expose the limits of rationality in human nature. Rational behavior is often defined as the pursuit of enlightened self-interest, whereas self-defeating behavior consists precisely in thwarting one's enlightened self-interest. Hence modern views of human nature as rational, information-processing beings often confront their limits in self-defeating behavior.

A decade ago, Baumeister and Scher (1988) reviewed a dozen major patterns of self-defeating behavior that social and personality psychologists had documented among normal (nonclinical) populations. By examining the set of them together, they were able to draw a series of broad conclusions. There was hardly any sign that people ever intentionally sought to suffer or fail. Rather, self-defeating behavior resulted either from tradeoffs, in which people pursued positive gains that ended up being accompanied by risks and costs, or they pursued positive outcomes but used strategies and methods that tended to backfire and produce unintended, undesired harm.

The role of emotional distress remained, however, as a major loose end in that work. Baumeister and Scher's (1988) review of the literature suggested that negative emotional states were involved in many of the self-defeating behavior patterns, but it was not clear how emotion produced such effects. In particular, there was little evidence to support psychodynamic hypotheses that negative affect makes people want to suffer, desire to fail, or believe they deserve to be punished. Why, then, should aversive emotional states increase self-defeating behaviors?

Risk-taking offered one possible answer that emerged from a review of the suicide research literature (Baumeister, 1990). Many deaths, such as from single-car crashes, are difficult to classify as either definite suicides or definite accidents. Some researchers have proposed that the difficulty of classification is more than a methodological limitation: it reflects the fact that the suicidal person did not have clear intentions either way. The person may have simply been taking extreme risks as one symptom of his or her highly distraught condition, and at some point the risks caught up with him or her.

To investigate whether risk-taking might mediate between emotional distress and self-defeating behavior, two of us (Leith & Baumeister, 1996) conducted a series of studies. One consisted of autobiographical narratives. We had people write accounts of their own past self-defeating actions. More precisely, they wrote about things they had done that had led to bad consequences that they later regretted, which seemed a fair operationalization of self-defeating behavior. For comparison purposes, they also wrote about something they had done that had turned out well. These stories were subjected to rigorous content coding for specific features relevant to our ideas.

The stories provided encouragement for the risk-taking theory. People were much more likely to describe taking risks or chances when they wrote about things that turned out badly than when writing about things that turned out well. Furthermore, there was a strong tendency for people to describe bad moods and emotional distress preceding the risk-taking. This was true in both types of stories — that is, even when they wrote about a risk that led to a positive outcome, they still tended to start the story with some bad mood or negative emotion that preceded the risky decision. In these accounts, bad moods were far more likely than good moods to lead to risk-taking.

We then developed a laboratory procedure to measure risk-taking, especially risks that seemed ill-advised or self-defeating. The measure, which we have used in a long series of studies (Leith & Baumeister, 1996, and subsequent unpublished work), involves presenting the subject with a choice between two lotteries. One of these is a low-risk, low-payoff lottery, offering a 70% chance of winning a $2 prize. The other is a long shot, with a $25 prize but only a 2% chance of winning.

Two features of this procedure need to be emphasized, given our interest in self-defeating behavior. First, although some people might count the lack of a positive outcome as bad, we thought there ought to be some actually negative outcome involved in order that the choice might qualify as self-defeating. Hence we added the stipulation that if one did not win the prize in the lottery, then one would be subjected to an unpleasant experience. The subject was told that this would involve taking part in a noise stress procedure, which meant putting on headphones in a sound laboratory and listening to a sound described as similar to fingernails scratching on the blackboard, magnified 25 times.

Second, it is important to recognize that the expected gains from the two lotteries are unequal. (The term *expected gain* is used by statisticians, accountants, and others to assess risks, and it is based on multiplying the probability of each outcome times the value of the outcome and then adding these up. For example, the low-risk lottery described above had an expected gain of .70 times $2 plus .30 times zero, or $1.40.) Even if one ignores the noise stress outcome and focuses solely on the cash payoffs, the expected gain from the low-risk lottery was nearly three times that of the long shot. The noise stress possibility further increased the discrepancy between the rational appeal of the two lotteries. Rational analysis or statistical calculation would therefore dictate always choosing the low-risk lottery. The long shot therefore qualified as a foolish risk.

In a series of experiments, then, we manipulated mood and emotional states and then assessed preference between the two lotteries. Our results repeatedly confirmed the view that emotional distress leads to foolish risk-taking. They also replicated earlier findings by Isen and her colleagues (Isen & Geva, 1987; Isen, Nygren, & Ashby, 1988; Isen & Patrick, 1983) indicating that pleasant, positive moods make people risk averse.

In one study, we created embarrassment by having people expect to sing a corny, difficult song without accompaniment while being stared at and tape recorded. These people showed a high preference for the long shot lottery, unlike neutral and positive mood subjects who expressed more evenly divided preferences among the two lotteries. In another, we created anger by asking people to recall and describe an intense interpersonal conflict and then by frustrating them with repeated equipment problems and requests for them to start over with their description. They too chose the long shot, while people in neutral and good moods showed a strong preference for the low-risk lottery. In yet another study, some people were put in a bad mood by jogging in place, and they too favored the long shot.

The only negative mood induction that failed to produce a preference for the long shot was sadness. We induced sadness in one group of participants by having them watch an excerpt from a sad movie. These people preferred the low-risk lottery just like the people in neutral and happy moods. Sadness differs from anger and embarrassment in several ways, but the most likely candidate is arousal.

The implication is that only high-arousal emotions produce the tendency to take foolish risks. Moreover, moods marked by high arousal and pleasant feelings did not yield that tendency either. Thus, self-destructive risk-taking appears to be concentrated in the mood and emotional states that combine unpleasant valence and high arousal.

But what does all this have to do with self-regulation? Initially we did not think there was any connection, beyond the simple point that self-defeating behavior can be defined as one type of poor self-regulation. Over the course of this investigation, however, we gradually began to realize that self-regulation plays a much more important and prominent role.

The initial findings confirmed the basic hypothesis that bad moods lead people to take stupid or foolish risks. But why? There is no obvious or direct connection between emotion and risk-taking. Thus, in a sense we had solved one problem only to reveal another. Risk-taking was apparently the link mediating between emotional distress and self-defeating behavior — but what was the link that mediated between emotional distress and risk-taking?

Our initial theory was that emotional distress alters the subjective utility of possible outcomes. In plainer terms, when you feel bad, you have more to gain and less to lose than when you feel good. The mirror image of this argument had been put forward by Isen et al. (1988) to explain why good moods made people risk averse: people do not want to take a chance on any outcome that might spoil their good mood.

By extension, risk-taking might be rationally appealing to someone in a bad mood. If one takes a chance and fails, one would feel bad, but insofar as one already felt bad, less is lost. Meanwhile, if one takes a chance and succeeds, one has not only the practical or material benefits of the success but also the improvement in one's mood, because the success would make one feel better.

Although this theory seemed plausible, we were not able to find any support for it. In study after study, people's ratings of the subjective appeal and value of the various outcomes showed no effect of their emotional states. Nor did their subjective estimates of the probabilities of winning or losing either of the lotteries change. Ultimately we had to abandon that theory.

The subjective value theory was based on the notion that emotion altered the cognitive appraisal of risk. An alternative hypothesis was that emotion simply cut short the cognitive appraisal of risk, rather than altering the values and calculations. To put it simply, maybe people who are upset simply grab for a desirable outcome without thinking about the dangers and consequences. This second hypothesis involved self-regulation, because it indicated that emotion prevented people from making an informed, considered choice and instead simply promoted impulsive choices. There was some precedent for this view: Keinan (1987) found that people who were in unpleasant, stressful situations tended not to review all the options on a multiple-choice test and instead simply selected the first viable one that presented itself.

To pit these theories against each other, we arranged to replicate the anger study with one further twist. In one condition, we instilled anger in participants and then gave them the lottery choice, but we told them to list the favorable and unfavorable features of each lottery before making their selection. We reasoned that if our first theory was correct and emotional distress altered the subjective utility calculations, then this effect would be magnified by calling people's attention to the relative merits of the two lotteries. In contrast, if our second theory was correct and emotional distress had its effect by preventing people from thinking through their options, then forcing them to think through those options should eliminate or reverse the pattern of results.

The results were quite clear: Angry people who had to stop and think for a few seconds before making their choice ceased to favor the long shot and instead overwhelmingly chose the low-risk lottery, just like people in neutral or positive moods. If stopping to think reverses the normal effect of anger, then presumably the normal effect of anger has to do with failing to stop and think. The implication is that states of emotional distress promote impulsive behavior that can produce risky and self-defeating outcomes. This is a form of self-regulation failure. In our procedure, when people did stop to think, they made the rational, optimal choice, but such analysis requires self-regulation to restrain oneself from making quick, impulsive choices. Emotional distress apparently undermines that self-regulation, thereby allowing people to make quicker choices without adequate consideration.

Together, these results provide a useful insight into the psychology of self-destruction. Emotional distress does indeed lead to self-defeating behavior, but the causal process does not necessarily conform to outdated psychodynamic theories about wishing for punishment or desiring to fail. Rather, emotional distress apparently undermines the self-regulation process necessary to produce an informed, rational appraisal of the situation and its options. Once that is gone, people tend to make impulsive choices without thinking them through. These choices may especially favor high-payoff but high-risk courses of action. Although sometimes the person may be lucky and benefit from the high payoff, in many other cases the risk will materialize into substantial harm or cost. The causal process thus leads from emotional distress (high arousal, negative moods) to self-regulation failure, to taking foolish risks, to costly self-defeating outcomes.

NATURE OF SELF-REGULATION

Our more recent line of research has addressed a basic question about the nature of self-regulation. What sort of process is involved in exerting self-control, as in resisting temptation, altering one's emotional state, or keeping a resolution?

At least three major types of theories have been described. One resembles the traditional concept of willpower, which implies that self-regulation involves energy or strength. Another is that self-regulation is essentially a matter of cognitive processing guided by a schema or knowledge structure. A third is that it is a skill that is acquired through practice. All of these are plausible and have their adherents (e.g., Baumeister et al., 1994; Carver & Scheier, 1981; Higgins, 1996).

The three theories make competing predictions about what will happen when people are presented with the need to self-regulate after they have just completed an act of self-regulation. If self-regulation is a strength, then it will become tired or depleted by the initial act, and so the second act is likely to be less effective. In contrast, if self-regulation is a schema, then the initial act will prime the schema (i.e., activate the relevant knowledge structures) and so subsequent self-regulation will be improved. And if self-regulation is a skill, then there would be no change in the second act, because skill remains essentially constant from one trial to the next.

Our first study to assess these competing predictions involved measuring the effects of affect regulation on physical stamina and endurance (Muraven, Tice, & Baumeister, in press). Participants watched a sad, upsetting movie excerpt with one of three instructions. Some were told to control their feelings so as to minimize any emotional response, including both trying to not show and not feel emotional reactions. Others were told to try to increase and amplify their emotional response (which is another form of self-regulation of emotion). A third group was not told to alter their emotional response in any way.

Following exposure to the film, physical endurance was measured by performance on a handgrip task. This task uses a device sold commercially for exercising the muscles in the hand, and it consists of two handles and a spring. The user squeezes the handles together, thereby compressing the spring. We measured how long people could maintain the handles squeezed together, which becomes physically tiring as the muscles use up their strength. Like any endurance task, a good performance requires one to use self-regulation to overcome the impulse to stop to rest, so that one can keep going.

The results of this first study supported the strength theory of self-regulation. People who had tried to regulate their emotions, either increasing or stifling their emotional response to the film, subsequently gave up faster on the handgrip task, as compared to people who watched the same movie but did not try to regulate their emotions. Thus, apparently, regulating one's emotional state used up some limited resource that was then unavailable for helping one persist on the handgrip task.

A second study used quite different procedures. In this one, people first engaged in a thought suppression task in which they try not to think about a white bear (borrowed from Wegner, Schneider, Carter, & White, 1987). Two control groups were also run, including one in which people were permitted (although not required) to think about a white bear, and one in which no bears were mentioned. Following this, we measured persistence on unsolvable problems. Each participant was asked to work on a series of anagrams, and these had been rendered unsolvable by omitting letters. The measure was how long the person kept trying to solve the puzzles before giving up.

This study likewise confirmed the strength or energy model. People who had tried to suppress thoughts about the white bear subsequently gave up faster on the anagrams task, as compared to people who had not tried to suppress forbidden thoughts. (Additional studies and control groups have ruled out alternative explanations, such as the idea that people felt themselves to have failed at the white bear suppression task and were therefore too discouraged to do well on the anagrams.) Thus, the act of trying not to think about the bear apparently depleted some crucial resource that was then unavailable for helping them persist at the problems in the face of failure.

Resisting temptation is one of the classic, prototype cases of self-regulation, and so we also conducted research with resisting temptation (Baumeister, Bratslavsky, Muraven, & Tice, in press). Participants in this study were asked to skip a meal prior to the experimental session, so most of them arrived hungry. To increase their temptation, we baked chocolate chip cookies in the laboratory room just before the subject arrived, so that the room and adjacent hallway were suffused with the delicious aroma of freshly baked chocolate. After entering the room, the subject was seated at a table near a stack of these cookies, along with chocolate candies. There was also a bowl of radishes on the table. In the crucial condition, the experimenter informed the subject that he or she was assigned to the radish condition and would have to eat only radishes. The subject was left alone in the room (to maximize the temptation to filch some chocolate) for five minutes.

Afterward, we measured how long people would continue trying to solve some geometric puzzles that had been rigged to be unsolvable, a measure borrowed from Glass and Singer (1972). The people who had had to resist the temptation gave up faster than people who had been allowed to eat chocolates, and also faster than people for whom no food had been involved. Thus, resisting temptation apparently depleted the same resource that was needed for persistence.

Next, we conducted several studies aimed at remedying some weaknesses or limitations in these first few. We were concerned that most of the task measures had included unsolvable tasks, and so we showed that an initial exertion of self-regulation led to a decrement in performance on solvable tasks (Baumeister et al., in press). We were also concerned that the dependent measures had all been oriented toward self-regulation of performance processes, so we conducted one in which the dependent measure was affect regulation. In that study, people first performed the white bear thought suppression exercise (or a neutral task not requiring self-regulation), and then they watched a funny movie with instructions to avoid laughing or smiling. Sure enough, the white bear exercise apparently depleted people's capacity for self-regulation, with the result that they were more likely to laugh and smile at the video.

All these results point to the conclusion that self-regulation conforms to a pattern of strength or energy. That is, there appears to be one inner psychological resource that is used for many different kinds of self-control, and any act of self-control that depletes the resource will result in subsequent decrements in self-control. In everyday life, this problem might correspond to people who find themselves confronted with exceptional demands for self-control and as a result begin finding that their self-control breaks down in other, unrelated spheres. For example, many students find that during examination week they are more likely to overeat, smoke too much, neglect their exercise, become crabby or show other signs of failed affect regulation, and in general exhibit various breakdowns in self-control. These may be understood on the basis of the limited energy model: final exam week requires the student to devote his or her self-regulatory capabilities to studying and complying with other academic deadlines and requirements, and this exceptional depletion of regulatory energy results in weaker self-control in other spheres.

One disturbing implication of this line of research is that it seems to warn people against exercising self-control. If people have only limited resources, then it seems prudent to avoid using it except when it is most urgently needed. People could take that as a recommendation to avoid exerting self-control, which would be a regrettable, socially undesirable recommendation.

A possible counterargument would be that if self-regulation is like a strength or muscle, then exercise should improve it over the long run. A muscle grows tired each time it is exercised but grows stronger as a result. It is conceivable that self-regulation would follow the same pattern.

We conducted one longitudinal study (Muraven, Baumeister, & Tice, 1997) to examine whether the capacity for self-regulation could be increased by exercise. We instructed several groups of students to follow some procedure for two weeks, and these procedures were all secretly designed as exercises in self-control. One group was told to try to improve their posture as often as possible. Another was supposed to try to regulate their moods and emotional states so as to feel good as much as possible. Two other groups were instructed to keep detailed records of everything they ate. A control group received no instructions.

At the beginning and the end of this two-week period, we assessed self-regulatory depletion by putting people through one of our experimental procedures. Specifically, they had to perform the white bear suppression exercise, followed by the handgrip measure of physical endurance.

The results of this study provided encouraging evidence that self-control can be improved with exercise. Relative to the control group, the groups who performed various self-control acts were less susceptible to depletion after the two weeks. That is, they showed improvement in how well they could make themselves persist on the handgrip task from the beginning to the end of the two-week period, whereas the control group did not show any such improvement (and in fact did worse, reflecting perhaps that the second measure was taken at a somewhat more stressful part of the semester).

Hence these results provide converging evidence that it is best to think of self-regulation as a kind of strength or muscle. Exertions of self-control bring short-term decrements but long-term improvements. Instead of concluding that people should avoid self-control, this line of work suggests that the optimal recommendation may be to exert self-control on a regular basis so as to increase one's overall capacity.

This line of research did lead into one more important set of findings that has potentially powerful implications. The studies covered thus far seemed to indicate that all (or at least a broad variety of) acts of self-regulation draw on the same pot of energy, which is quite limited in size. That pot of energy would have to be regarded as an important part of the self, given the importance of self-regulation. Still, is that pot used only for self-regulation?

A review of what social psychologists have learned about the self yielded the conclusion that relatively little is known about the "executive function" in general (Baumeister, in press). The executive function is another term for what is sometimes called the agent or the active principle (and is an important part of what is called the "I"). More precisely, it is the part of the self that initiates action, makes choices and decisions, and takes responsibility.

Self-regulation can be regarded as one of the duties of the executive function. If self-regulation uses this limited supply of energy, could it be that the other duties of the executive function also draw on that same resource? If so, that resource would indeed be one of the most important and central aspects of the self.

To examine whether the same energy resource is used for activities outside of self-regulation, we needed to show that other acts would interfere with subsequent self-regulation. For our first study, we used the standard choice manipulation from cognitive dissonance research (cf. Linder, Cooper & Jones, 1967). In this procedure, subjects were asked to make a counterattitudinal speech favoring a large tuition increase at their university. Half were simply told to make the speech (using materials that were given them and being recorded on audiotape). Others were asked to make the speech but told that the final decision about whether to do so was entirely up to them. Following this, we used the same persistence measure that we had used in the radish/chocolate study.

The results provided striking confirmation of the view that choice draws on the same energy resource that self-regulation does. People who had performed the act of choice and taken responsibility for their behavior acted like depleted subjects in the previous studies: they gave up relatively quickly on the unsolvable problems. In contrast, people who had performed the same counterattitudinal behavior without choice showed no decrement in persistence, as compared to the no-speech control group.

In other work, we have found that initial acts of self-control make people more passive. Apparently active responses by the self draw on the same energy resource that self-regulation does. When people face a decision where there are both active and passive options, they are more likely to take the passive option if they have previously exerted self-control.

Taken together, these results suggest implications about the basic nature of the self, and not just about self-regulation. The self seems to have a limited supply of energy that is required for a variety of its functions, including active responses (e.g., initiating action), making choices and decisions, taking responsibility, and regulating the self. It appears that the same resource is used in all these and, moreover, that the resource is quite easily depleted.

CONCLUSION

Self-regulation has major, important implications for success in life, and indeed there is ample basis for asserting that it is one of the most important keys to success. People who are good at self-regulation show a multitude of advantages over other people, in both task performance and interpersonal relations.

This chapter has provided an overview of our recent studies on self-regulation. This work has indicated that self-regulation failures are involved in a variety of self-defeating behavior patterns, including taking foolish, destructive risks and procrastinating.

The studies on risk-taking indicate that high-risk behavior may often be an important link between emotional distress and self-defeating behavior. When people are upset (in the form of unpleasant moods or emotional states that feature high arousal), they tend to cut short the processing of information about the possible costs and bad outcomes associated with some choices. They may pursue a highly desirable goal but one that carries risks and costs that would make it prohibitive if one were to consider the options thoughtfully. Self-regulation is required to make oneself stop and think instead of acting impulsively. Emotional distress apparently undermines this form of self-regulation.

Procrastination involves postponing work on assigned tasks. Contrary to apologists who regard procrastination as fairly harmless, the present results suggest that it carries significant costs in terms of task performance and health. In a series of studies, we found procrastinators earned lower grades, suffered more stress, and were sicker than nonprocrastinators. The only benefit of procrastination was that when the deadline was far off they enjoyed lower stress and better health than nonprocrastinators — gains that were later offset and even reversed when the deadline became close. Self-regulation includes the self-discipline to make oneself work on tasks in the absence of powerful external pressures, and so self-regulation is inimical to procrastination.

An important form of self-regulation involves self-management, which refers to using information about self and the external world to choose appropriate performance settings and other commitments so as to maximize what one can accomplish. We found that people with high self-esteem were better at this when operating under favorable conditions. An ego threat, however, disrupted the smooth self-management of people with high self-esteem and caused them to become overly confident, resulting in destructive and costly courses of action.

Our last line of work addressed the fundamental question of the nature of self-regulation. Converging evidence from many studies suggests that self-regulation operates like a stock of common energy, analogous to a muscle or other form of strength. When people exert self-regulation, their capacity becomes depleted, similar to the way a muscle becomes fatigued after exercise. For a time, any other act of self-regulation is likely to show poorer performance, as long as the fatigue or depletion lasts. Meanwhile, preliminary findings indicate that self-regulation can improve over time with regular exercise, which extends the muscle analogy.

The findings about depletion suggest that the capacity for self-regulation depends on a limited supply of energy, and this supply must be regarded as one of the most important features of the self. Apparently the same stock of energy is used for a wide variety of different acts of self-regulation, including many that would seem entirely unrelated to each other. Furthermore, recent findings suggest that this supply of energy is also used by the self for all acts of choice, initiative, responsible decision-making, and active response. This supply of energy is thus vital to the effective functioning of the self.

Yet the supply is also quite limited. A mere five minutes of resisting temptation to eat chocolate, for example, led to a subsequent reduction by half in the amount of time people could persist in the face of failure before quitting. Whether the self's energy is really absent in such conditions, or whether the effects indicate only that the self responds to depletion by seeking to conserve, it is still clear that the self can only exert deliberate control over a very small amount of its behavior. The functioning of the self must therefore be understood as taking place amid scarcity, and the reliance on habits, routines, automatic behaviors, and easy or passive patterns of action may be a direct result. Fortunately, however, current work suggests that the self's capacity for self-regulation can be increased.

REFERENCES

Baumeister, R. F. (1990). Suicide as escape from self. *Psychological Review, 97*, 90–113.

Baumeister, R. F. (1997a). Esteem threat, self-regulatory breakdown, and emotional distress as factors in self-defeating behavior. *Review of General Psychology, 1*, 145–174.

Baumeister, R. F. (1997b). *Evil: Inside human violence and cruelty*. New York: W.H. Freeman.

Baumeister, R. F. (in press). The self. In D. T. Gilbert, S. T. Fiske, & G. Lindzey (Eds.), *Handbook of social psychology* (4th ed.). New York: McGraw-Hill.

Baumeister, R. F., Bratslavsky, E., Muraven, M., & Tice, D. M. (in press). Ego depletion: Is the active self a limited resource? *Journal of Personality and Social Psychology*.

Baumeister, R. F., Heatherton, T. F., & Tice, D. M. (1993). When ego threats lead to self-regulation failure: Negative consequences of high self-esteem. *Journal of Personality and Social Psychology, 64*, 141–156.

Baumeister, R. F., Heatherton, T. F., & Tice, D. M. (1994). *Losing control: How and why people fail at self-regulation*. San Diego, CA: Academic Press.

Baumeister, R. F., & Scher, S. J. (1988). Self-defeating behavior patterns among normal individuals: Review and analysis of common self-destructive tendencies. *Psychological Bulletin, 104*, 3–22.

Baumeister, R. F., Smart, L., & Boden, J. M. (1996). Relation of threatened egotism to violence and aggression: The dark side of high self-esteem. *Psychological Review, 103*, 5–33.

Berglas, S. C., & Baumeister, R. F. (1993). *Your own worst enemy: Understanding the paradox of self-defeating behavior*. New York: Basic Books.

Boice, R. (1989). Procrastination, busyness, and bingeing. *Behavior Research and Therapy, 27*, 605–611.

Boice, R. (1996). Binge writing and procrastination/blocking amongst new professors. Manuscript submitted for publication.

Campbell, J. D. (1990). Self-esteem and clarity of the self-concept. *Journal of Personality and Social Psychology, 59*, 538–549.

Campbell, J. D., & Lavallee, L. F. (1993). Who am I? The role of self-concept confusion in understanding the behavior of people with low self-esteem. In R. Baumeister (Ed.), *Self-esteem: The puzzle of low self-regard* (pp. 3–20). New York: Plenum.

Carver, C. S., & Scheier, M. F. (1981). *Attention and self-regulation: A control theory approach to human behavior*. New York: Springer-Verlag.

Ferrari, J. R. (1992). Psychometric validation of two adult measures of procrastination: Arousal and avoidance measures. *Journal of Psychopathology & Behavioral Assessment, 14,* 97–100.

Ferrari, J. R., Johnson, J. L., & McCown, W. G. (1995). *Procrastination and task avoidance: Theory, research, and treatment.* New York: Plenum

Glass, D. C., & Singer, J. E. (1972). *Urban stress: Experiments on noise and social stressors.* New York: Academic Press.

Gollwitzer, P., & Bargh, J. A. (1996). *The psychology of action.* New York: Guilford.

Gottfredson, M. R., & Hirschi, T. (1990). *A general theory of crime.* Stanford, CA: Stanford University Press.

Higgins, E. T. (1996). The "self digest": Self-knowledge serving self-regulatory functions. *Journal of Personality and Social Psychology, 71,* 1062–1083.

Isen, A. M., & Geva, N. (1987). The influence of positive affect on acceptable level of risk and thoughts about losing: The person with a large canoe has a large worry. *Organizational Behavior and Human Decision Processes, 39,* 145–154.

Isen, A. M., Nygren, T. E., & Ashby, F. G. (1988). Influence of positive affect on the subjective utility of gains and losses: It is just not worth the risk. *Journal of Personality and Social Psychology, 55,* 710–717.

Isen, A. M., & Patrick, R. (1983). The effect of positive feelings on risk-taking: When the chips are down. *Organizational Behavior and Human Performance, 31,* 194–202.

Keinan, G. (1987). Decision making under stress: Scanning of alternatives under controllable and uncontrollable threats. *Journal of Personality and Social Psychology, 52,* 639–644.

Lay, C. H. (1986). At last, my research article on procrastination. *Journal of Research in Personality, 20,* 474–495.

Lay, C. H. (1995). Trait procrastination, agitation, dejection, and self-discrepency. In J. R. Ferrari, J. L. Johnson, & W. G. McCown' (Eds.), *Procrastination and task avoidance: Theory, research, and treatment* (pp. 97–112). New York: Plenum.

Leith, K. P., & Baumeister, R. F. (1996). Why do bad moods increase self-defeating behavior? Emotion, risk-taking, and self-regulation. *Journal of Personality and Social Psychology, 71,* 1250–1267.

Linder, D. E., Cooper, J., & Jones, E. E. (1967). Decision freedom as a determinant of the role of incentive magnitude in attitude change. *Journal of Personality and Social Psychology, 6,* 245–254.

McCown, W., & Johnson, J. (1989, April). Validation of an adult inventory of procrastination. Paper presented at the *Society for Personality Assessment,* New York.

McCown, W., & Johnson, J. (1991). Personality and chronic procrastination by students during an academic examination period. *Personality and Individual Differences, 12,* 662–667.

McFarlin, D. B., & Blascovich, J. (1981). Effects of self-esteem and performance feedback on future affective preferences and cognitive expectations. *Journal of Personality and Social Psychology, 40,* 521–531.

Mischel, W. (1974). Processes in delay of gratification. In L. Berkowitz (Ed.), *Advances in experimental social psychology* (Vol 7, pp. 249–292). San Diego, CA: Academic Press.

Mischel, W. (1996). From good intentions to willpower. In P. Gollwitzer & J. Bargh (Eds.), *The psychology of action* (pp. 197–218). New York: Guilford.

Mischel, W., Shoda, Y., & Peake, P. K. (1988). The nature of adolescent competencies predicted by preschool delay of gratification. *Journal of Personality and Social Psychology, 54,* 687–696.

Mischel, W., Shoda, Y., & Rodriguez, M. L. (1989). Delay of gratification in children. *Science, 244,* 933–938.

Muraven, M., Baumeister, R. F., & Tice, D. M. (1997). Longitudinal improvement of self-regulation through practice: Building self-control strength through repeated exercise. Manuscript submitted for publication.

Muraven, M., Tice, D. M., & Baumeister, R. F. (in press). Self-control as limited resource: Regulatory depletion patterns. *Journal of Personality and Social Psychology.*

Platt, J. (1973). Social traps. *American Psychologist, 28,* 641–651.

Shoda, Y., Mischel, W., & Peake, P. K. (1990). Predicting adolescent cognitive and self-regulatory competencies from preschool delay of gratification: Identifying diagnostic conditions. *Developmental Psychology, 26,* 978–986.

Taylor, S. E., & Brown, J. D. (1988). Illusion and well-being: A social psychological perspective on mental health. *Psychological Bulletin, 103,* 193–210.

Tice, D. M., & Baumeister, R. F. (in press). Longitudinal study of procrastination, performance, stress, and health: The costs and benefits of dawdling. *Psychological science.*

Vallone, R. P., Griffin, D. W., Lin, S., & Ross, L. (1990). Overconfident prediction of future actions and outcomes by self and others. *Journal of Personality and Social Psychology, 58,* 582–592.

Wegner, D. M., Schneider, D., Carter, S. R., & White, T. L. (1987). Paradoxical effects of thought suppression. *Journal of Personality and Social Psychology, 53,* 5–13.

9

IMPROVING COMPETENCE IN PRIME ADULTHOOD

Of Sows' Ears and Pygmalions

Alex E. Schwartzman

Centre for Research in Human Development
Department of Psychology
Concordia University

Developmental studies of competence tap into core areas of the life sciences concerning the nature, conditions, and constraints of organismic change. Work in these fields deals with the properties of homeostatic regulation, plasticity, and resource allocation in relation to growth, stability, and decline of function over the course of the human life cycle. In this chapter, we review the competence-related demands, deficits, and remedial needs of prime adulthood in this context, and we examine the remedial implications of current trends in theory and research in the cognitive and socioemotional domains of adult development.

When we speak of competence, we are, of course, dealing with a broad-band, multifaceted construct. Depending on the context, the concept of competence involves at least four points of reference: (a) standards of attainment or efficacy; (b) the idea of mastery, expertise, success, or stamina; (c) areas or contexts of challenge (physical, cognitive, social, emotional); and (d) sources of influence (maturation, experience, internal state, external conditions). When we speak of prime adulthood, we are referring to the age period roughly between 25 and 50 years that is stereotypically depicted as the time when many of our strengths and skills reach their peak and stabilize; when the biologically rooted properties of plasticity and growth begin to fade; when developmental continuity finds expression in the harvesting and consolidation of skills acquired in childhood and adolescence; and when questions of conservation and selective rationing of resources are not as yet a major concern.

Prime adulthood is also defined by the culture's developmental assignments. Havighurst's (1953) representation of these tasks almost half a century ago remains relevant despite the changes that have ensued in gender role expectancies, family constellation, and occupational patterns. The early phase includes partner selection, family life preparation, establishment of an economic standard of living, and the cultivation of social

Improving Competence across the Lifespan, edited by Pushkar *et al.*
Plenum Press, New York, 1998.

ties beyond family and work. The middle segment is directed primarily at child-rearing and economic advancement balanced by leisure-time activity. The latter years are devoted to the maintenance of affiliative ties and lifestyle, to age-appropriate parenting of adolescent children on the brink of adulthood, and to meeting the needs of aging parents.

Clearly, there is expansion and diversification of role-related competencies over the course of the prime-of-life years. The challenge at this juncture of the life span is to assimilate the new "hats" efficaciously and to relinquish those that are no longer relevant. As in other phases of development, there are stresses associated with these changes, the effects of which heighten the potential for maladaptive behavior and the need for remedial intervention. Any attempt to redress problematic adjustment, however, must contend with the developmental issues of plasticity, stability, and continuity in maturity. The stereotypic perception of prime adulthood as a period of consolidation and stability makes for pessimistic prognostications about the possible effectiveness of remediation at this stage of the life course, particularly if it is a matter of modifying well-entrenched maladaptive behaviors.

Theoretical formulations and evidence of recent vintage, however, bring into question a number of long-held tenets of the life sciences that have fueled conventional views of development. The principle of ontogenetic recapitulation of phylogeny with its implication of a hierarchical development of competencies in the human has come under critical scrutiny (see Gottlieb, 1992), as has the related notion of an orderly progression within competence domains from the immature, simple, and inefficient to the mature, complex, and efficient (see Bjorkland, 1997; Gould, 1977). In effect, the refashioned perspectives of development currently gaining currency afford a more nuanced and dynamic representation of plasticity constraints and remedial modifiability in prime adulthood.

The history of life span studies of psychometric intelligence is instructive in this respect. Advances in method and theory have unhinged mistaken or simplistic assumptions about age-related changes in intelligence. Culture-fair, experience-relevant, and age-appropriate measurement in a longitudinal, repeated-measures research format plus the application of a testing-the-limits paradigm (Kliegl, Smith, & Baltes, 1989) have generated findings that attest to the pluralistic nature of intelligence, the specificity and complexity of its component parts in terms of experience- and age-related gains and losses, and perhaps most importantly in the present context, the capacity of the aging individual to benefit to some extent from specific deficit-targeted remedial training programs (Schaie & Willis, 1986; see chapter by Dolores Pushkar and Tannis Arbuckle where the limitations of benefits to such programs are noted).

Evidence consistent with a multidimensional, multidirectional causal model of competence development, however, has come largely from research on the young and the elderly. Studies of adult competence in the years between early and late maturity have not as a rule been cast in developmental terms. A literature search is likely to foster the impression that students of development regard this stage of the life span as the asymptotal pause between growth and decline. A transactional, pluralistic perspective, however — one which moves away from a linear, reductionistic view — allows for developmental discontinuities and periods of transition that are prompted by major life events including those that encompass maturational milestones. In the present context, the question is whether the prototypical transitional periods of prime adulthood serve as windows of opportunity for the remediation of maladaptive, competence-related behaviors; and in broader terms, whether there are grounds to qualify conventional notions of a decreasing capacity for change with the passing years of maturity.

In the framework of Havighurst's (1953) developmental tasks, the move from single to partnered or family head status calls for work-related skills that assure attainment of a livable economic standard and for socioemotional competencies that make for meaningful and durable emotional attachments. Although deficits in the two competence domains are not unrelated, the penalties of poverty compel social policy planners to give first priority to vocational skill needs as one target of a multi-pronged corrective effort. The challenge for the applied social scientist is twofold: (a) to test the premise that motivation and amelioration of vocational skill deficits are associated with the imminence of change from single to first-time partnered or family head status; and (b) to design remedial strategies that relate meaningfully to the lives of low-income individuals. The role of context in cognitive development is relevant in this regard.

There is now accruing a line of evidence that argues for context and process specificity as key contributors in the development of cognitive competencies (see Ceci, 1993; Detterman, 1992; Hoffman & Nead, 1983; Johnson-Laird, 1983; Wilding & Valentine, 1988). Although it is recognized that there are capacity constraints of bio/experiential origin, failure to solve a problem in one context does not necessarily preclude successful solution of the same problem in another (Ceci, 1990). In addition, research demonstrating that heritability estimates of behavioral phenotypes can vary as a function of context (Bronfenbrenner & Ceci, 1993) provides support for the thesis that modifications of context alter genetic as well as environmental contributions. Simply put, context appears to be embedded in the very processes that code informational input. To the extent that context limits the generalization or transfer of skills, as this literature would suggest, there are opportunities to develop innovative remediation strategies that minimize the constraining effects of biology and negative learning experiences.

The goal, then, is to design remedial interventions that are perceived by the targeted constituency as meaningful and beneficial. To achieve this objective requires that context be manipulated as an integral part of a skills assessment protocol for low income groups. At least five avenues of contextual influence on cognitive performance can be tapped. First, physical contextual features can be varied on gradients of familiarity and valence. Warranting assessment, for example, is the effect of remedial settings other than the conventional classroom. Second, the influence of the interpersonal dimension on performance may be determined as a function of the size and sociodemographic attributes of the group (e.g., gender composition, single parent status), instructor characteristics (gender, age, style of teaching), and social support structures (e.g., "buddy" systems, counseling). Third, the mode of presentation of a cognitive task has been shown to be an important source of performance variation (e.g., video game, paper and pencil, stimulus configuration; Ceci, 1993). Fourth, assessment of the economy's current and long-term skill needs provides the motivational/historic context required to mount remedial programs that are realistically geared to marketable competencies. Fifth and finally, it is useful to manipulate or "unstandardize" the contextual constraints embedded in the standard administration of age-normed cognitive tests. Limit-testing procedures (e.g., speed limit variation, trial repetition to criterion) are particularly instructive and relevant in designing specific, deficit-targeted remedial strategies. They provide ball-park estimates of cognitive capacity reserves and constraints, and in so doing, enable the helping professional to tailor remedial goals accordingly. To summarize, models of cognitive development and methodological advances in the past three decades argue for specificity — specificity in the identification of cognitive processing deficits in prime adulthood, in the selection and framing of age-appropriate and context-relevant remedial strategies, and in the targeting of meaningful goals and criteria of goal attainment.

Applying the principle of remedial specificity to the socioemotional domain has proven difficult when the problematic competencies implicate broad-spectrum, maladaptive behavior patterns of long standing. The thrust of theory and evidence to date places emphasis on the primacy of biology, temperament, and early experience as the defining sources of influence on socioemotional development. Erikson (1980), in his life stage formulations, pointed to intimacy and generativity as the adult's prototypical socioemotional goals, the realization of which depended largely on the levels of socioemotional competence attained in earlier stages of development; and prospective studies that have followed children into maturity attest to the continuity and stability of normative and maladaptive socioemotional behavior patterns (e.g., Caspi, Bem, & Elder, 1989; Farrington, Gallagher, Morley, Ledger, & West, 1988; Robins, 1978; Rutter, 1987; Schwartzman, 1992).

Findings of this kind reaffirm the notion of a "sow's ear" principle at work. It should be noted, however, that there is an age-related gradient of controllability of life experience (see Greenough, Black, & Wallace, 1987, on "experience-expectant" and "experience-dependent" brain plasticity; also Heckhausen, 1997, on primary and secondary control) and an age-related modulation of uncontrollable genetic input (Thompson, 1993). Taken together, the two sources of influence allow for a "pygmalion" potential.

Accordingly, the current challenge for applied social scientists is to determine the conditions in prime adulthood that optimize the potential for change on the dimensions of socioemotional competence — attachment capacity, emotional reactivity, valence, intensity, and behavioral tilt (approach/avoidance). Specifically, there is a need for research which tests the validity of a three-pronged proposition to the effect that under conditions of socioemotional challenge, (a) there remains a measurable range of plasticity and reserve capacity in the resources that impinge on the dimensions of socioemotional competence; (b) access to these resources is under voluntary control; and (c) there are temporal (e.g., times of role transition), task-specific, and situational contexts which enhance the probability of exercising voluntary control to achieve adaptive change.

The proposition calls for theory and evidence relating behavior to process, process to category, and category to time frame in prime adulthood. The theoretical impetus of research on socioemotional development has until recently been category-oriented. External and internal behavior referencing of emotional states (e.g., Ekman, Levenson, & Friesen, 1983; Izard, 1991), developmental stage formulation (Erikson, 1980), and person traits (Costa & McCrae, 1990) have served to structure the field's areas of research. For the applied social scientist dealing with the competence deficits of prime adulthood, categorization is informative to the extent that it bears on context-specific processes; process-oriented study is informative to the extent that it bears on manipulable conditions and mechanisms of change; and manipulation is informative to the extent that it bears on the strategies selected to modify maladaptive performance of the major tasks of prime adulthood: the initiation and maintenance of nurturant spousal and parent–child relationships, modulating planful, adaptive modes of goal-directed behavior, and achieving an enduring sense of well-being over the course of the mature years.

In category-driven research, longitudinal studies which follow children at psychosocial risk into prime adulthood provide the opportunity to examine the correlates of three types of developmental trajectory: (a) the "resilient" profile refers to those who in the face of adversity, remain competent and well-functioning (Rutter, 1990); (b) the "recovered" label applies to emotionally troubled children who "grow out" of their difficulties (Rutter, 1990); and (c) the "sleeper" pattern pertains to originally competent children who develop problems by prime adulthood. This line of research informs the remedial literature. It provides a developmental overview of the conditions and coping strategies associated with adaptive and maladaptive changes.

Process-oriented study is usually associated with the reframing of categorydirected theoretical perspectives. Emotional expression defined by emotional state, for example, has fostered category-related research (e.g., Izard, 1991) Emotional expression defined as social signalling has spurred interest in age-related process variables that modulate the salience and reciprocal effects of parent–child socioaffective communication (see Thompson & Leger, 1994).

Emotional attachment framed in a "separation" theoretical perspective has generated category-related studies of correlates of "secure/insecure" status (see Lamb et al., 1985). Attachment viewed as a socioemotional representational system has prompted interest in age-related processes of internal "working models" — models of social relatedness that pertain to the self in relation to early and later attachment figures (Bretherton, 1990). Clearly, this line of research has particular relevance in considering questions of attachment capacity in the development of intimate affiliative ties. Needed in this area are longitudinal assessments of the processes influencing change in the internal representations of spousal and parent–child relationships over the course of prime adulthood.

In a similar vein, the classification of parent authority styles (cf. Baumrind, 1971) has given rise to an extensive category-oriented research literature. At the same time, it has served as an entree to theoretical models and evidence which deal with the age-related and family context-specific processes which modulate parental expressions of authority in adaptive and maladaptive ways (cf. Dix, 1991; Hetherington & Clingempeel, 1992; Patterson & Yoerger, 1993; Steinberg, Elmer, & Mounts, 1989). The means by which inimical belief systems and negative emotion in parenting behavior can be modified remains one of the major priorities of remedial research in this area (cf. Goodnow & Collins, 1990; Kochanska, 1991; Maccoby, 1992; Zahn-Waxler & Kochanska, 1990).

Adaptive modes of emotional control are clearly a requisite feature of socioemotional competence above and beyond the realm of parenting behavior. Studies in this area as in those mentioned earlier have been category-oriented. The focus has been on individual differences in temperament attributes (easy, difficult, inhibited, excitable) as biologically rooted sources of influence on the regulation of emotional arousal over the course of the life cycle (see Kagan, 1992).

Temperament as a construct serves the same organizing function in the socioemotional developmental literature as does intelligence in the cognitive domain. It has also, like intelligence, proved difficult to define. Possibly for this reason, students of socioemotional development appear to be following the same path of research as those in the cognitive domain initiated several decades ago. The challenge for developmentalists in the socioemotional domain at this juncture is to formulate developmental models which generate process-oriented studies of emotional arousal which ask the question, how it affects the self and others (see Eisenberg & Fabes, 1994; Garber & Dodge, 1991).

Thompson (1993, p. 379) defines emotional regulation as "the extrinsic and intrinsic processes responsible for monitoring, evaluating, and modifying emotional reactions, especially their intensive and temporal features, to accomplish one's goals." In this theoretical frame, the social dimensions of emotional arousal are brought to the fore. They are key features of the regulatory processes of emotion implicated in the minutia of family life and goal-directed activity in prime adulthood.

Stresses and strains are an integral part of the minutia, and the types of coping behavior they trigger are an integral feature of the processes regulating emotional arousal. Despite the definitional vagaries of stress as a construct, our understanding of its impact on emotional arousal and control has been advanced by process-oriented studies which examine coping behavior as a function of person, stressor, and context-specific characteristics (see Aldwin, 1994). The challenge in this area is to examine further in a developmental context, the prototypical sources of stress, their history and significance for the individual, and their effects. As Bruce Compas demonstrates in his chapter, there are clear differences between children, adolescents, and adults in how they perceive and deal adaptively and maladaptively with life-threatening stressors.

To summarize, research on competence deficits in the cognitive domain has advanced sufficiently to generate innovative and process-targeted remedial programs for individuals who are about to assume the responsiblities of prime adulthood. Although considerably more research is needed in the area of socioemotional development to reach the same objective, there has been sufficient progress made in the last decade to warrant an optimistic outlook.

REFERENCES

Aldwin, C. M. (1994). *Stress, coping, and development*. New York:Guilford.

Baumrind, D. (1971). Current patterns of parental authority. *Developmental Psychology, 4* (Monograph 1), 1–103.

Bjorkland, D. F. (1997). The role of immaturity in human development. *Psychological Bulletin, 122,* 153–169.

Bretherton, I. (1990). Open communication and internal working models: Their role in the development of attachment relationships. In R. A. Thompson (Ed.), *Socioemotional development: Nebraska symposium on motivation 1988* (Vol. 36, pp. 59–113). Lincoln, NE: University of Nebraska Press.

Bronfenbrenner, U., & Ceci, S. J. (1993). Heredity, environment, and the question "how?": A first approximation. In R. Plomin & G. McClearn (Eds.), *Nature, nurture, and psychology* (pp. 313–325). Washington, DC: American Psychological Association.

Caspi, A., Bem, D. J., & Elder, G. H. (1989). Continuities and consequences of interactional styles across the life course. *Journal of Personality, 157,* 375–406.

Ceci, S. J. (1990). On the relationship between microlevel and macrolevel cognitive processes: Worries over current reductionism. *Intelligence, 14,* 1–19.

Ceci, S. J. (1993). Contextual trends in intellectual development. *Developmental Review, 13,* 403–435.

Costa, P. T., Jr., & McCrae, R. R. (1990). Personality and the five-factor model of personality. *Journal of Personality Disorders, 4,* 362–371.

Detterman, D. K. (1992). Transfer on trial: The case for the prosecution. In D. K. Detterman & R. J. Sternberg (Eds.), *Transfer on trial: Intelligence, cognition, and instruction.* Norwood, NJ: Ablex.

Dix, T. (1991). The affective organization of parenting: Adaptive and maladaptive processes. *Psychological Bulletin, 110,* 3–25.

Eisenberg, J., & Fabes, R. A. (1994). Emotion, self-regulation, and social competence. In M. Clark (Ed.), *Review of personality and social psychology.* Newbury Park, CA: Sage.

Ekman, P., Levenson, R. W., & Friesen, W. V. (1983). Autonomic nervous system activity distinguishes between emotions. *Science, 221,* 1208–1210.

Erikson, E. H. (1980). *Identity and the life cycle.* New York: Norton. (Original work published 1959)

Farrington, D. P., Gallagher, B., Morley, L., Ledger, R. J., & West, D. J. (1988). Are there any successful men from criminogenic backgrounds? *Psychiatry, 51,* 116–130.

Garber, J., & Dodge, K. A. (Eds.). (1991). *The development of emotion regulation and dysregulation.* Cambridge: Cambridge University Press.

Gottlieb, G. (1992). *Individual development and evolution: The genesis of novel behavior.* New York: Oxford University Press.

Goodnow, J. J., & Collins, W. A. (1990). *Development according to parents.* Hillsdale, NJ: Erlbaum.

Gould, S. J. (1977). *Ontogeny and phylogeny.* Cambridge MA: Harvard University Press.

Greenough, W., Black, J., & Wallace, C. (1987). Experience and brain development. *Child Development, 58,* 539–559.

Havighurst, R. J. (1953). *Human development and education.* New York: McKay.

Heckhausen, J. (1997). Developmental regulation across adulthood: Primary and secondary control of age-related challenges. *Developmental Psychology, 33,* 176–187.

Hetherington, E. M., & Clingempeel, W. G. (1992). Coping with marital transitions: A family systems perspective. *Monographs of the Society for Research in Child Development, 57,* (Serial No. 2–3).

Hoffman, R. R., & Nead, J. M. (1983). General contextualism, ecological science and cognitive research. *The Journal of Mind and Behavior, 4,* 507–560.

Izard, C. E. (1991). *The psychology of emotions.* New York: Plenum.

Johnson-Laird, P. N. (1983). *Mental models: Toward a cognitive science of language, inference, and consciousness.* Cambridge, MA: Harvard University Press.

Kagan, J. (1992). Yesterday's premises, tomorrow's promises. *Developmental Psychology, 28,* 990–997.

Kliegl, R., Smith, J., & Baltes, P. B. (1989). Testing the limits and the study of adult age differences in cognitive plasticity of a mnemonic skill. *Developmental Psychology, 25,* 247–256.

Kochanska, G. (1991). Socialization and temperament in the development of guilt and conscience. *Child Development, 62,* 1379–1392.

Lamb, M. E., Thompson, R. A., Gardner, W., & Charnov, E. L. (1985). *Infant-mother attachment.* Hillsdale, NJ: Erlbaum.

Maccoby, E. E. (1992). The role of parents in the socialization of children: An historical overview. *Developmental Psychology, 24,* 1006–1017.

Patterson, G. R., & Yoerger, K. (1993). Developmental models for delinquent behavior. In S. Hodgins (Ed.), *Mental disorder and crime* (pp. 140–172). London, UK: Sage.

Robins, L. (1978). Sturdy childhood predictors of adult outcome: Replications from longitudinal studies. *Psychological Medicine, 8,* 611–622.

Rutter, M. (1987). Continuities and discontinuities from infancy. In J. Osofsky (Ed.), *Handbook of infant development* (2nd ed., pp. 1256–1296). New York: John Wiley & Sons.

Rutter, M. (1990). Psychosocial resilience and protective mechanisms. In J. Rolf, A. S. Masten, D. Cicchetti, K. H. Neuchterlein, & S. Weintraub (Eds.), *Risk and protective factors in the development of psychopathology* (pp. 181–214). New York: Cambridge University Press.

Schaie, K. W., & Willis, S. L. (1986). Can decline in adult intellectual functioning be reversed? *Developmental Psychology, 22,* 223–232.

Schwartzman, A. E. (1992). *Individual differences in childhood aggression and social withdrawal as predictors.* Quebec Symposium on Childhood and the Family, Quebec, QC.

Steinberg, L., Elmer, J. D., & Mounts, N. S. (1989). Authoritative parenting, psychosocial maturity, and academic success among adolescents. *Child Development, 60,* 1424–1436.

Thompson, R. A. (1993). Socioemotional development: Enduring issues and new challenges. *Developmental Review, 13,* 372–402.

Thompson, R. A., & Leger, D. W. (1994). From squalls to calls: The cry as a developing socioemotional signal. In B. Lester, J. Newman, & F. Pedersen (Eds.), *Biological and social aspects of infant crying.* New York: Plenum.

Wilding, J., & Valentine, E. (1988). Searching for superior memories. In M. M. Gruneberg, P. Morris, & P. Sykes (Eds.), *Practical aspects of memory* (Vol. 2). London: Wiley.

Zahn-Waxler, C., & Kochanska, G. (1990). The origins of guilt. In R. A. Thompson (Ed.), *Socioemotional development: Nebraska Symposium on Motivation* (Vol. 36, pp. 183–258). Lincoln, NE: University of Nebraska Press.

10

RESPONDING TO THE CHALLENGES OF LATE LIFE

Strategies for Maintaining and Enhancing Competence

Sherrie Bieman-Copland,[1] Ellen Bouchard Ryan,[2] and Jane Cassano[3]

[1]Department of Psychology
Brock University
[2]Department of Psychiatry and Office of Gerontological Studies
McMaster University
[3]Department of Psychology
McMaster University

The People were nomads living in the far north. They survived by moving in search of food. The winter had been a particularly harsh one for the People. Two elderly women in the group, Ch'idzigyaak and Sa', had become burdensome because of their slow pace and complaints. Worried about the survival of the group the People faced a desperate decision, whether or not to abandon the two old women. There was much disagreement among the People. While abandonment was not uncommon in bleak times, Ch'idzigyaak and Sa' seemed too young and capable for such a practice. The usual practice would have been to abandon the women with nothing so that they would die in the night. However, because of the feelings of uncertainty and guilt among the People, Ch'idzigyaak and Sa' were left with a tent, a hatchet, and a bundle of babiche, which are thick strips of moose hide used for many things.

Ch'idzigyaak and Sa' were hurt and angered by the actions of the group. They prepared to die as they settled in their tent for the night. Much to their surprise, they awoke the next morning and were very much alive. Sa' became hopeful and suggested to her companion that if they were to die, they may as well die trying to survive. Together, the two women remembered how to set traps with the babiche and they caught a squirrel with the hatchet that was left behind. They quickly realized that game around the campsite was too scarce, and they knew they had to move on if they were to make it through the winter. Ch'idzigyaak told Sa' an old story of a Place the People had visited in her youth which had an abundance of fish and other game. With nothing to lose, they decided to make their way towards this fruitful spot which was many days journey away. The traveling was long and tiresome. The women walked for many hours each day with little rest. They would awake in the morning feeling great aches and pains in their bodies. Several days into their journey they realized they had left behind the walking sticks that they had needed when they had traveled with the People. Their old complaints now seemed trivial, and they felt remorse about the way they had behaved with the People.

Eventually they made it to the Place and the fish and game were as plentiful as Ch'idzigyaak had remembered. In the Place Ch'izigyaak and Sa' flourished. They survived the win-

Improving Competence across the Lifespan, edited by Pushkar *et al.*
Plenum Press, New York, 1998.

ter, and when summer came they worked hard preserving animals and making warm clothing for the following winter. At the end of the summer, they had more food and clothing than the two of them needed.

(Abridged from Wallis, 1994)

"The Tale" highlights a major theme for this chapter, that is, the social facilitation of the non-use of competence. As the two women aged, changes in abilities likely occurred, the types of changes which have been well documented in the cognitive and biological aging literature (Finch & Schneider, 1985; Salthouse, 1991). However, a more important factor in shaping their behavior was the reduced expectations from the tribe (i.e., age stereotypes) of what they could contribute, and a progressive loss of role and status. Over time, the tribe, and even the women themselves, became more aware of their limitations and frailties. Unfortunately, the outward behavior of the women (e.g., slow pace, complaining), served only to reinforce the tribe's beliefs and behaviors.

The lifespan developmental approach provides a useful framework for understanding the process of aging, and identifies mechanisms which contribute to successful aging. From this perspective changes which occur in late development are not solely considered to be a result of chronological age or age-related declines in biological processes but the result of multiple social, psychological and biological influences (Baltes & Baltes, 1990). The ability to maintain mastery over simple tasks (e.g., self-care) and more complex tasks (e.g., managing finances) is viewed as a contributor to an older adult's level of perceived competence, and is also interpreted as a indicator of cognitive ability (Baltes, Mayr, Borchelts, Maas, & Wilms, 1993; Willis, 1996). Performance on formal evaluations, which largely focus on cognitive ability, has also been used as a way to measure older adults' levels of competence. However, as we discuss throughout the chapter, formal evaluations may be problematic and, therefore, may not accurately reflect the older adult's level of competence.

In this chapter, we describe a lifespan framework of successful aging which includes three primary adaptive mechanisms, namely selection, optimization, and compensation (Baltes & Baltes, 1990; Marsiske, Lang, Baltes, & Baltes, 1996). We also discuss the influence of non-aging factors on development. Interventions that have enabled older adults to overcome competence predicaments will be reviewed. These interventions primarily targeted either the older individual, the task at hand, or the social/physical environment in such ways as to break the negative feedback cycle described so eloquently in "The Tale."

THE COMPETENCE PREDICAMENT: THE SOCIAL FACILITATION OF THE NON-USE OF COMPETENCE

Negative expectations about aging have their basis in negative stereotypes about old age in terms of poor memory, frailty, increasing dependency and inflexibility (Hummert, 1990; Kite & Johnson, 1988) and in terms of social policies that limit roles for older adults which lead to reduced social status and prestige (Palmore, 1990). Through a process of socialization, young adults internalize the stereotypes prevalent in society and enter old age with negative expectations about aging. As a result, older adults behave in ways that are thought to be typical and characteristic of their age group (Heckhausen & Lang, 1996). Older adults who experience declines that are consistent with age stereotypes may not try to change their behavior because they believe it represents the norm. In a similar vein, stereotyped expectations also lead older adults to accept the diminished role society gives them and they behave "old" (Rodin & Langer, 1980).

Age stereotypes also bias how the performance of older adults is interpreted by observers. An identical memory, achievement, or communication performance will be interpreted differently depending on whether the target is young or old. Erber and her colleagues demonstrated the existence of an age-based double standard in memory appraisals. She found that people were more likely to interpret benign, everyday memory failures as a sign of mental impairment when they occurred in old compared to young adults. Such memory failures were seen to be caused more by lack of ability (a non-controllable source) when they occurred in old targets but caused by lack of effort (a controllable source) when they occurred in young targets (e.g., Erber, 1989; Erber & Rothberg, 1991; Erber, Szuchman, & Rothberg, 1990a, 1990b; also Parr & Siegert, 1993). In addition, Bieman-Copland and Ryan (in press) demonstrated that such age-biased interpretations were robust even when salient situational variables, which could account for the memory outcome, were included in the scenarios. Successes and achievements are also interpreted in a stereotypical manner. For example, Bieman-Copland and Ryan found that everyday memory successes were judged to be less typical for old compared to young targets. Likewise, Ryan, Szechtman, and Bodkin (1992) found that young participants rated old targets as less likely than young targets to succeed in completing a computer course despite giving the old target who was taking the course higher competence ratings than the young target. Ryan and Laurie (1990) found that when messages were communicated by old versus young men, the messages were evaluated less positively and perceptions of the men who communicated them were also less positive. Moreover, young listeners failed to differentiate good versus poor communication performances of old but not young speakers.

The Communication Predicament of Aging model was developed to highlight how these stereotyped expectations and interpretations influence intergenerational communication and contribute to age-related losses in functioning (Ryan, Giles, Bartolucci, & Henwood, 1986; Ryan, Hummert, & Boich, 1995). According to the framework, conversational partners sometimes base their speech with older adults on erroneous assumptions guided by beliefs about decline and incompetence in old age. Such age-biased modifications can include high-pitch, exaggerated intonation, slow pace, simplification, repetition, reduced politeness, terms of endearment, and talk about the past. While some degree of modification may be necessary for some older individuals, these patronizing styles of speaking are elicited solely on the perception of general old age cues (e.g., white hair, stooped posture) rather than on the needs of the particular individual. These speech modifications reinforce age-stereotyped behaviors such as dependence and inactivity, and subsequently create barriers to positive interaction. In the extreme, continued exposure to such situations may eventually lead to reduced self-esteem in the older adult, and withdrawal by young and old communicators from intergenerational exchanges.

Aging stereotypes and expectations can also influence an older adult's level of perceived control. Frail elders are more often in low control environments such as nursing homes or hospitals (Teitelman & Priddy, 1988), which are typically socially and physically restrictive. Although staff intend to be helpful, their behaviors are often based on stereotypes about older adults. Research in nursing home environments found that staff engaged more frequently in actions which caused further dependency among residents (Baltes & Wahl, 1992). Seligman (1975) and other researchers have developed the theory of learned helplessness and found that individuals who are frequently exposed to low control physical and/or social environments may experience several deficits. There are declines in motivation, cognitive ability, as well as emotional disturbances which are particularly consequential for frail older adults. A common response to being exposed to low control environments is passive behavior which is usually interpreted by others as a sign of incompetence.

The function of stereotypes may be so strong that it can also affect formal evaluations of competence. Too often old age is considered the primary contributor to poor performance in older adults and multiple factors which influence performance are frequently overlooked. A competence enhancing approach to evaluation and intervention involves appropriate consideration of the multiple influences (Ryan, Kwong See, Meneer, & Trovato, 1992; Willis, 1991).

Cohort effects rather than age per se may explain some age differences in performance. Evaluations that are completed by psychologists and others often occur in school-like settings and involve unfamiliar tasks which put older adults at a disadvantage because of their lack of recent experience. For example, the education of older adults differs both quantitatively (e.g., number of years of formal schooling) and qualitatively (e.g., teaching techniques used) from younger adults. This situation is aggravated if there are arbitrary time limits for completing the task. For older adults, it is critical that tasks which evaluate competence have high levels of face validity (i.e., be similar to tasks they perform in everyday life). An older adult may be poorly motivated to complete tasks that lack relevance for them. Fatigue may also adversely affect performance.

Individual life histories provide important information about the life long skills and abilities which may influence how current performance is evaluated. For example, an older adult may have always experienced difficulty with memory in a particular domain such as remembering names. One older adult we interviewed about perceptions of her memory ability replied, "I take heart in knowing that what I forget now are things I've always had difficulty remembering" (Bieman-Copland & Ryan, 1996). Life time experiences tend to magnify individual differences among older adults in areas such as world knowledge, domain specific expertise, and information processing strengths and weaknesses.

Recent experience with health changes, levels of activity, and environmental stimulation may also markedly affect a person's performance at a given time (Lawton, 1982; Rowe & Kahn, 1987). An older person may have to be more conscientious about the time of day and the physical environment (e.g., background noise, poor lighting, glare) in which they perform activities if optimal levels of competence are to be demonstrated. Furthermore, one must also be cognizant of the fact that older adults may set different task goals affecting their views of what is meaningful and what is an ideal strategy (Brigham & Pressley, 1988). Likewise, subtle biases within examiners may be inadvertently communicated to older adults further reducing their expectation and motivation to exert effort.

Rowe and Kahn (1987) differentiate usual aging from successful aging based on the degree to which non-aging or extrinsic factors are controlled. They argue that many of the effects that the scientific literature has described as being caused by age, are in fact due to a greater degree by factors other than age. The term successful aging has become the "buzz" word of gerontological researchers in the 1990s. Successful aging is often defined in terms of exceptional performance or achievements by older adults (Baltes & Carstensen, 1996), an end state which may be unachievable by a majority of older adults.

A LIFESPAN DEVELOPMENTAL FRAMEWORK FOR SUCCESSFUL AGING

In contrast, a lifespan developmental framework emphasizes that successful aging is a process rather than an end state. Figure 1 displays a modified version of concepts critical for this perspective which were introduced by Baltes and Baltes (1990). Successful aging is a highly individualized and subjective concept and is recognized when an older individ-

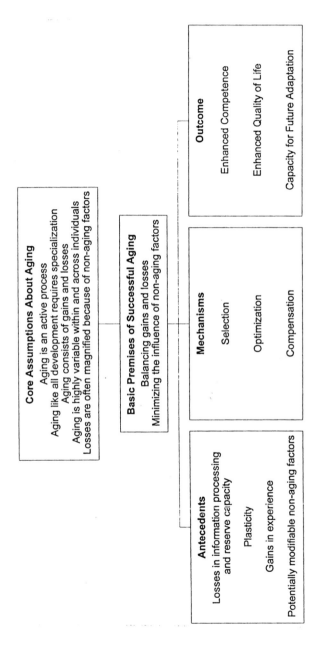

Figure 1. A lifespan developmental framework of successful aging. [Adapted from Baltes & Baltes, 1993.]

ual is able to achieve desired goals with dignity and as independently as possible. This means that successful aging is possible for a nursing home resident whose goal is to be able to maintain health and complete basic activities of daily living as well as for a recent retiree whose goal is to be an active member in the community.

Core Assumptions about Aging

Aging like all development involves processes of specialization. In aging, part of this specialization involves increasing knowledge acquired through life experiences. But as with any specialization, there is a cost in that potential expertise in areas not chosen is sacrificed. Because opportunites for experience are so vast and individual, variation both within and across individuals tends to increase with age. This heterogeneity is the hallmark of late adulthood.

For older adults, the process of specialization occurs in the context of clear age-related changes in physiological capacities. In the past, the focus in adult developmental research has largely been on losses in domains such as sensory perception, information processing, and memory. Because of this focus, late adulthood has been generally accepted as a time of accumulating losses and is equated with declining competence. While losses may outweigh gains, the possibility of maintaining or increasing adaptive capacity is a timely concept which the Baltes approach has brought to the study of human development. The key for adaptive functioning is ensuring that older adults work with rather than against changes in physiological capacity. This notion of development as an active and dynamic process replaces the more simplistic developmental notion of childhood as a period of growth while aging is a period of decline.

We propose an extension of the Baltes model that considers factors unrelated to age which affect the performance of older adults. We believe that the focus on losses within adult developmental research has lead to a situation where such factors have been largely ignored. The distinction between age differences and age changes is often blurred and, as a result, losses attributed to aging per se are magnified. What is unfortunate about this situation is that many factors which adversely affect the performance of older adults are modifiable, and appropriate interventions could lead to increased competence.

Basic Premises of Successful Aging

In contrast to normal aging, the process of successful aging is influenced by a combination of factors which are internal to the individual, and external in the social or physical environment. An internal factor considered to be a basic premise of successful aging is maintaining a balance between gains and losses which occur simultaneously at any point across the lifespan. Although, the number of losses experienced in late life may outweigh the number of gains, there is the possibility of positive developmental change during the latter part of the lifespan. This balance can be achieved through the use of various mechanisms which will be described in more detail later in the chapter. These processes allow older adults to adapt and adjust to changing abilities and environments, characteristic of late adulthood.

The external factors, which are unrelated to the aging process, need to be addressed as these may affect an older adult's perceived level of competence and successful aging. Older adults' use of existing skills or their ability to acquire new ones to perform a given task can often be challenged by non-aging factors which are most often found in the environment. As discussed earlier on, factors such as perceptual environmental conditions, low expectation, or lack of meaningful task materials may lead to negative outcomes. As well, non-aging factors are compounded with age-related changes (e.g., poor vision) which serve to accentuate poor performance, but are not direct reflections of true ability. The influence of such factors may unnecessarily affect an older adult's level of perceived competence, and their beliefs about their ability to master events in the environment.

Antecedents

In recent years a major focus in the field of cognitive aging has been to identify the age-related declines in memory, language, and other cognitive domains (for extensive reviews, see Blanchard-Fields & Hess, 1996; Craik & Salthouse, 1992; Kausler, 1994). A strong position within the cognitive aging literature has emerged, suggesting that a wide range of findings can be accounted for on the basis of changes in some of the underlying elementary processes, such as speed of information processing or working memory (Park, et al., 1996; Salthouse, 1991). In other words, at this point in time, there is a general consensus that aging causes significant declines in information processing and reserve capacity. At the same time, there is enormous discrepancy between high levels of everyday functioning in older adults and that which would be predicted based on laboratory studies (Salthouse, 1987). We believe that since the focus on age declines is consistent with stereotyped expectations of late adulthood as a period of loss, researchers have not adequately examined the factors (such as, accumulated background knowledge and 'know-how') which older adults use to maintain functional competence.

The study of older experts tells us a great deal about how experience may or may not offset age-related losses. In a classic study, Charness (1981a, 1981b) selected young and old chess players with similar levels of expertise based on their competitive ratings. He found no significant differences between the two groups on standardized measures developed to assess skill level (i.e., selecting the next move and evaluating game outcomes). In contrast, older chess players were less accurate in a task where they were required to recall the positions of chess pieces placed in meaningful configurations on a board. In a similar vein, Salthouse (1984; see also Bosman, 1993) found that older typists were able to maintain typing speed while still displaying the expected age-related deficits in laboratory reaction time tasks. This means that life time experience did not protect the older adults from the expected age-related changes in basic information processing domains. However, such changes did not necessarily negatively affect their competence in areas of expertise.

Other studies have demonstrated that experience may allow some attenuation of age-related differences. For example, older adults compared to younger adults have more difficulty acquiring basic computer skills (Kelley & Charness, 1995). However, Kelley, Charness, Mottram, and Bosman (1994) found no age difference in measures reflecting learning a new computer application among individuals who already had computer skills. Studies of age-related changes in verbal memory ability have shown smaller age differences in individuals with high versus low levels of verbal ability (Cavanaugh, 1990; Hultsch, Hertzog, & Dixon, 1990; Meyer & Rice, 1989) or in individuals with active versus passive lifestyles (Craik, Byrd, & Swanson, 1987). Such individual differences somehow protect older adults from some of the memory declines usually associated with age, however, the mechanism by which such protection occurs has not been articulated.

Salthouse and Somberg (1982) demonstrated that experience derived from practice within a laboratory setting may be sufficient to reduce age-related differences. These authors found large initial age differences in performance on simple detection, discrimination, and speeded classification tasks. However, practice improved the performance of both young and old participants, with improvement being "perhaps even greater in older subjects than in young subjects" (p. 201). In other words, task experience (consisting of 51 practice sessions) enhanced the performance of old adults as much or more than it did for young adults. Moreover, this improvement was equally evident in both age groups over a period of one month. Other studies have shown similar effects of practice with shorter periods of intervention in a broader range of mental abilities (see Schaie & Willis, 1986). At the same time, one should not expect that practice, even extensive practice, on one type of mental ability task will lead to enhanced performance on another. That is, cognitive retraining tends to have specific rather than generalized effects.

In a similar vein, we have found it interesting that studies which demonstrate the phenomenal degree of plasticity within cognitive structures and processes of older adults are interpreted in the literature strictly in terms of losses in reserve capacity. In a series of studies, researchers in Berlin (Kliegl, Smith, & Baltes, 1986, 1989, 1990) taught young and old participants a complex strategy for learning word or number sequences, then provided them with extended opportunities for practice. They found that the old participants, like the young participants were able to use these strategies to make extraordinary gains in digit span. Without practice or strategy instruction, digit span performance for both young and old adults was approximately seven (plus or minus two) digits. With extensive practice, but without the strategy training, the span performance of neither young nor old participants exceeded sixteen digits. However, given both extensive practice and specific strategy training, digit span performance increased to over 100 digits for individuals in both age groups. When this research is cited in the cognitive aging literature it is usually used to demonstrate that age differences become magnified at extreme levels of performance because old adults are only able to demonstrate this type of extraordinary performances when the presentation of digits occurred at a very slow rate (i.e., one digit every 20 seconds), whereas young adults can achieve this level of performance over a much wider range of task conditions. However, we see this as a demonstration of the extraordinary potential that is present within older adults.

Potentially modifiable non-aging factors may unnecessarily impinge upon older adults' perceived or measured capacity in a number of domains. Many of the standards selected by researchers to judge performance are often inherently youth biased. This may be the case whenever age differences are examined on measures that are time-based or unpracticed. For example, early research in aging and intelligence concluded that aging led to dramatic declines in fluid intelligence abilities. However, this research likely overestimated these declines because the standardized administration of such tests dictated that they be performed under strict time constraints, used materials that were novel, and may have been unduly influenced by fatigue (Botwinick, 1984). Therefore, there was an inherent confounding of speed of performance, task familiarity, and fatigue with the construct of interest (i.e., fluid intelligence). While more recent research continues to show that fluid abilities decline with age, such changes are less dramatic than once thought (Berg & Klaczynski, 1996).

Mechanisms

In the following section we describe and provide examples of three particular mechanisms that can lead to preserved or enhanced levels of everyday functioning despite losses in some domains. Clear theoretical distinctions are made between these mechanisms of selection, optimization and compensation and we talk about them as separate strategies. However, the reader should be cognizant that in reality all three strategies tend to operate together and continuously interact.

Selection. Selection is a process by which older adults readjust goals to ensure that they are able to achieve acceptable levels of performance, within a narrower range of activities. This means that older adults may drop certain activities to focus on those areas which they perceive to be most important, where abilities and skills are most preserved and/or where interest is highest. Such selection highlights the process of specialization which characterizes adult development, serving to maintain or enhance the competence of older adults.

An important part of selection is cognitive restructuring, whereby older adults perceive selected activities to be of greater importance and devalue those goals which are not selected. Through such cognitive activity older adults maintain a sense of personal control. We have seen evidence of such control strategies in our research on memory beliefs of older adults (Bieman-Copland & Ryan, 1996). In this study, we asked older subjects to describe how they perceived their memory abilities had changed with age. Some participants acknowledged that their memory ability had declined with age, but interpreted such change as positive rather than negative (e.g., "Forgetting occurs more frequently now because my sense of values have changed. Lots of things that were once important to me I no longer consider worth remembering."). Other participants contended that their memory had improved with age because of associated positive age-related changes (e.g., "I would say my memory is better now because I've learned to use my memory over the years and have more confidence in myself.").

Selection mechanisms may influence strategy preference in memory-demanding situations. Brigham and Pressley (1988) had young and old adults learn unfamiliar words with one of two mnemonic strategies. Although both appeared to be reasonable learning strategies, one led to more effective recall. After one learning trial, a majority of young but not old participants stated a preference for the more effective method. Questioning after the experiment suggested that the preference decisions by young participants were guided almost entirely by recall enhancement, while older participants tended to base their preference decisions on familiarity and meaningfulness. Difference in strategy selection can also be seen in the study of old and young chess players. Charness (1985) found that while older chess players selected a move as soon as they found one that was acceptable, younger players continued to search for better alternatives until a time limit was reached.

In a social domain, Carstensen (1991) examined patterns of interaction and social exchange throughout the lifespan. As expected, she found that older adults had a more limited social network. However, rather than being an inevitable consequence of aging, she found that older adults were active agents in the process and participants generally viewed the quality of their social network quite positively. Highly valued social contacts tend to remain stable across the lifespan, while less highly valued contacts change slowly from mid-life on, and then dramatically in old age.

An implication of selectivity for researchers is that goals which researchers deem optimal (e.g., high accuracy in recall, large social networks) may not be viewed in the same way by older adults. On the negative side, the process of selection may lead older adults to prematurely adjust goals meaning that areas where competence could be achieved are not pursued. Individuals are at risk for setting goals based on their chronological age as opposed to their ability and skill (Dittman-Kohli, 1990). For example, some older adults may have the potential to achieve high goals such as obtaining a university degree, however they have been shaped to believe that selecting such a goal is inappropriate. In short, premature goal adjustment may be one consequence of the negative feedback loop described earlier, and result in the non-use of well preserved abilities.

Optimization. The mechanism of optimization involves the maintenance and enhancement of existing reserves or resources so that optimal levels of functioning are achieved (Baltes & Carstensen, 1996). Optimization serves to protect the skills and abilities needed to perform highly valued and selected goals in the face of age-related declines. This is partly achieved through the selection of appropriate environments that serve to promote or enhance performance (e.g., choosing appropriate living arrangements or social partners, performing tasks in selected goal areas under optimal physical conditions), as well as through learning new skills or expanding existing knowledge bases. Practice is a key feature of optimization. As an individual becomes more familiar with a particular task and the necessary skills involved, their efficiency at performing the task increases because of increasing automaticity. Studies which have demonstrated optimization have been previously discussed in the context of plasiticity.

Compensation. Compensation is a process whereby older adults change the way they perform tasks in order to minimize the mismatch between personal resources and contextual demands (Backman & Dixon, 1992; Dixon & Backman, 1995). This mismatch may arise because of individual deficits, caused by aging or non-aging factors, which means the individual is unable to maintain competence despite no change in environmental demand. Alternatively, the mismatch may arise because of an increase in environmental demands that is not matched by an increase in the individual's abilities or skills (Backman & Dixon, 1992; Dixon & Backman, 1995). The advantage of compensation is that older adults may be able to continue engaging in valued activities that they might have otherwise abandoned. The disadvantage of the process is that they no longer complete the activity in the usual manner or as efficiently as previously. An individual may find such a situation to be unacceptable and prefer to abandon the activity. Cognitive restructuring interventions may be needed to make the modified activity more acceptable to the individual.

Compensation includes (a) modifying skills used to perform a target task, (b) transferring skills from other domains that are relevant to the present task but not previously used, or (c) collaborating with others so that the full range of skills required to complete a task are available to the "team." Examples of compensatory behavior in older adults are prevalent in the psychological aging literature. In the driving domain, older adults compared to younger adults drive fewer miles per year, drive less frequently on the highway, at night, or during rush hour (Charness & Bosman, 1995). In the domain of memory, research has shown that compared to young adults, older adults make greater use of external memory aids relative to internal mnemonic strategies (Cavanaugh, Grady & Perlmutter, 1983). In the skill domain, Salthouse (1984; also see Bosman, 1993) found that older typists were able to produce transcription typing speeds equivalent to younger typists, but did so by increasing the number of characters in their preview span (i.e., the number of characters by which their visual scanning exceeded their motor response). This work is important because it articulated a mechanism by which older adults' experience allowed them to compensate for age-related declines in speed of motor responses. Moreover, it provided an example where compensation occurred without awareness. Therefore, contrary to the position of Backman and Dixon (1992), compensation does not necessarily involve a conscious recognition of the mismatch between ability and task requirements.

Defining Successful Aging Outcomes

The processes of selection, optimization and compensation are particularly beneficial in light of the various social, physical and environmental changes experienced by older adults. The ability to perform tasks which are highly valued and maintain mastery over everyday activities may enhance an older adult's level of perceived competence

(Heckhausen & Lang, 1996). Associated with increases in perceived competence are feelings of self-esteem, self-worth, control and independence. These self perceptions also contribute to physical health and well-being (Ryff, 1989). Whereas "The Tale" described a negative feedback loop, the processes involved in successful aging promote a positive feedback loop, an enhanced quality of life and a greater capacity for future adaptation.

INTERVENTION RESEARCH DIRECTED TOWARDS IMPROVING COMPETENCE IN OLDER ADULTS

In this section, we will review selected research studies which have attempted to preserve and increase the abilities of older adults. The primary goal of many of these interventions was to bring performance levels of older adults up to those displayed by young adults who receive no intervention (i.e., attempts to eliminate age differences). We will argue that such a goal may not be the most appropriate one for intervention research. From a practical point of view, interventions must expand the focus beyond the older individual to include modifications to the task and physical/social environment.

Interventions Directed toward the Older Adult

One of the most active areas of intervention research in the elderly has been in improving memory. Studies of memory have consistently shown that older adults tend to encode information less actively and make less use of memory strategies to enhance retrieval (see Craik, 1977; Poon, 1985). Therefore, it would seem reasonable that providing instruction in specific mnemonic strategies might be useful in improving the memory competence of older adults. Mnemonic techniques such as the method of loci, the pegword method, face-name, and keyword strategies have been successful with young adults (see Searleman & Herrmann, 1994). Although their application to older adults have met with mixed results, a meta-analytic study by Verhaeghen, Marcoen, and Goossens (1992) reported that pre-post test gains in memory performance were significantly larger in groups receiving training than in control or placebo groups. Unfortunately, there is also evidence suggesting that older adults who learn the memory strategies do not actually use the strategy later on, even when tested in similar circumstances (e.g., Anschutz, Camp, Markley, & Kramer, 1987; Robertson-Tchabo, Hausman, & Arenberg, 1976). West (1989) cites a number of reasons for such lack of maintenance. For example, the strategies that are taught may have been too resource demanding, and therefore, yield too high a cost for the benefits they accrue in memory performance. Finally, strategies may not be taught in a step-by-step manner that would lead to better procedural consolidation of the strategy.

However, an alternate explanation is that the goals of the researchers (i.e., to improve memory performance) may have been quite different from the goals of the older adults (i.e., to feel better about their memory ability). Scogin, Strorandt, and Lott (1985) developed a memory training protocol which resulted in improved memory performance, but did not reduce memory complaints. Lachman, Weaver, Bandura, Elliot and Lewkowicz (1992) developed a multi-faceted memory intervention intended to both improve memory performance and reduce memory complaints. Their approach to memory training was unique in that those in the memory training condition were provided with information about different approaches to enhancing memory but they were required to individually develop their own memory strategies based on this intervention. Those in the memory beliefs conditions received information designed to change how they thought about the role of effort in memory functioning so as to alter beliefs that all memory failures in old age are inevitable and uncontrollable. Interestingly, all treatment groups including the no treatment group improved from pre-test to post-test. However, those who received both individualized memory training and belief restructuring were more likely to believe that they could improve their memory. Unlike other training studies, those in the memory training conditions reported three months after training that they were still using the strategies learned in training to help them remember in everyday situations.

Intervention studies have also focused on helping older adults develop new skills. Much of the skill development research has focused on teaching older adults to use word processing and other types of common computer software. This research has demonstrated that older adults, even those with no previous computer expertise can learn computer skills. However, they learn less efficiently than and differently from comparable novice young adults. For example, they require about twice the amount of time to move through tutorial lessons and they make more requests for assistance from the instructor during learning (Elias, Elias, Robbins, & Gage, 1987; Czaja, Hammond, Blascovich, & Swede, 1986; Zandri & Charness, 1989). Both young and old adults make greater gains when learning is collaborative (i.e., with a partner) than when taught individually, even though it decreases the amount of time each person is in direct contact with the computer (Zandri & Charness, 1989). Such collaborative learning may be more effective because it facilitates active problem solving when errors occur. For older adults, training may need to specifi-cally address interference effects that arise because the similarity in appearance between a computer keyboard and a typewriter may cause them to engage knowledge that actually interferes with acquiring computer skills.

Other interventions geared toward the individual are more familiar and have been pro-moted for many years. Research on the effects of engaging in regular physical activity sug-gests a positive relationship between exercise and functional capacity in a variety of domains including cognitive tasks. For the aging body, participation in physical activity improves cardiovascular functioning, flexibility, muscle strength, and reaction time (Fletcher & Hirdes, 1996). Studies also indicate that active older adults maintain a higher level of psychological functioning and well-being compared to older adults who are inactive (Mihalko & McAuley, 1996). With respect to cognitive ability, DiPietro, Seeman, Merrill, and Berkman (1996) found a modest but significant relationship between the ability to per-form cognitive tasks and the amount of physical activity reported by their subjects; how-ever, the component of exercise which affects cognitive functioning remains unclear. There is also a correlation between depressive symptoms and lack of physical activity (O'Connor, Aenchbacher & Dishman, 1993). The social aspect of physical exercise may be a mediating variable in the relationship between physical activity and psychological functioning.

Interventions aimed at promoting skills and abilities of older adults have shown good success across many domains. The studies discussed above demonstrate that older adults can be active participants in counterbalancing the negative changes associated with age. However, intervention needs to go beyond changing the older individual. The compe-tence of older adults can also be maintained or enhanced by altering tasks and physical and social environments.

Interventions Directed toward the Task and the Physical/Social Environment

To increase the accessibility of computers to older adults, modifications in computer design and machine–user interfaces need to be made. For example, the speed at which young adults can read from a computer monitor does not vary with the display color, but older adults read significantly faster with black on white displays (which most closely approximates reading from a printed page) compared to blue on black (Charness & Bosman, 1990). Similarly, using a light pen instead of a mouse interface reduced age differences in movement times on a computer pointing task and allowed older adults to perform difficult tasks such as moving to small targets (Charness, 1995). The 'double click' on the mouse interface is also problematic for older adults. To overcome this difficulty, Kelley and Charness (1995) suggested that the default double-click speed be reset for older adults. There is great potential for computers to enhance the competence of the older adult. More research into elder-friendly computer systems to enable this potential to be realized seems warranted.

The physical environment can be altered to fit the physical and social needs of older adults, thereby offsetting age-related declines. Environmental design interventions described by Hiatt (1987) create an environment which assists instead of hinders those with cognitive difficulty. Examples of such design interventions include minimizing of problem stimuli (e.g., glare), and reducing sensory overload (e.g., background noise), or selecting furnishings that facilitate physical activity and social interaction. It has been found that environmental adaptations especially in institutional settings can lead to an improvement in attention span, decrease agitation and may even alleviate problematic behavior of residents (Hiatt, 1987).

A number of effective intervention studies were conducted in an effort to demonstrate that modifications in social environments result in positive changes in physical health and well-being in institutionalized older adults. Langer and Rodin's (1976) initial study conducted in a nursing home implemented small measures (i.e., residents received a plant to care for) to increase the older individual's sense of personal responsibility and control in the nursing home. The results of this study showed that there were significant differences in self reports of well-being between the experimental and control groups and differences were still apparent on follow-up 18 months later (Langer & Rodin, 1976; Rodin & Langer, 1977). A similar study by Schulz (1976) suggested that perceived control was the critical element in the intervention. In his study, participants in the experimental group were given control over where and when social interactions with a new social partner would occur. The changes in psychological status and activity level that were found in these studies suggest that manipulation of the social environment of older adults can help reverse a pattern of progressive decline. These studies are some of the most influential in the aging literature because of the remarkable results obtained with relatively minor, inexpensive interventions.

More recently, Baltes and her colleagues have identified dependence-support scripts in interactions between nursing home staff and residents (Baltes & Wahl, 1992) and developed an intervention which targeted changing staff behaviors rather than that of the resident. In particular, staff behavior was altered so that they would react in a more reinforcing manner to independent behaviors displayed by the residents. Over time, the residents engaged more frequently in independent behaviors. The most important aspect of the study is that some of the onus for enhancing competence in older adults needs to be placed on those who interact with them.

SUMMARY AND CONCLUSIONS

This chapter reviewed many of the important conceptual and applied notions of aging. The intervention studies described are only a sample of those which have had a common goal of promoting and enhancing the competence of older adults. Effective interventions share several common characteristics. First, they are multifaceted. They attempt to make changes in the individual by facilitating processes of selection, optimization, and compensation, and in doing so make the older individual an active participant in the intervention. They examine the task and facilitate changes that would ensure a better match between task demands and abilities of the older adult. Finally, they examine the environment within which older adults perform tasks and make alterations which would enhance performance. Environmental modifications must go beyond the physical environment and consider the ways in which social interaction between older adults and others can lead to a non-use of competence. Sometimes interventions have been put in place which have not resulted in enhanced performance of older adults in everyday situations. This may result because the goals of older adults and researchers are often different. As researchers, we must be sensitive to how age stereotypes influence our own perceptions of older adults and remember that heterogeneity is the hallmark of old age. Because individuals grow old in diverse ways, interventions must be flexible. The role of psychologists should be to empower and treat older adults as unique individuals who do not share common needs. Some guidelines for how to communicate effectively so that older adults can enhance competence are offered in the communication enhancement model (Ryan, Meredith, MacLean, & Orange, 1995). The balancing act for psychologists involves adjusting to individual older persons' losses while maintaining appropriately high expectations and taking full advantage of the gains.

> While Ch'izigyaak and Sa' flourished in the Place, The People had not fared so well and when the next winter came they were in the same state that they had been in the year before. With feelings of guilt the People returned to where they had abandoned the two old women. They were surprised to find no sign of the women at the old campsite and the Chief sent out trackers to find the women. The People were surprised that the women were alive and were astonished at their present living condition. With new found independence and respect, Ch'idzigyaak and Sa' rejoined their People and helped the group survive a second harsh winter.

> (Abridged from Wallis, V., 1994)

The irony in "The Tale" is that had the knowledge and competence of the two old women been recognized and promoted by the tribe, a great deal of suffering within the tribe would have been avoided. So beyond enhancing the quality of life, interventions which promote competence have the potential to positively change our views of older adults from being a societal burden to a valuable resource.

REFERENCES

Anschutz, L., Camp, C. J., Markley, R. P., & Kramer, J. J. (1987). Remembering memories: A three year follow up on the effects of mnemonics training in elderly adults. *Experimental Aging Research*, *13*, 141–143.

Backman, L., & Dixon, R. A. (1992). Psychological compensation: A theoretical framework. *Psychological Bulletin*, *112*, 259–283.

Baltes, P. B. (1993). The aging mind: Potential and limits. *Gerontologist*, *33*, 580–594.

Baltes, P. B., & Baltes, M. M. (1990). Psychological perspectives on successful aging: The model of selective optimization with compensation. In P.B. Baltes & M.M. Baltes (Eds.), *Successful aging: Perspectives from the behavioral sciences* (pp. 1–27). New York: Cambridge University Press.

Baltes, M. M., & Carstensen, L. L. (1996). The process of successful ageing. *Aging and Society*, *16*, 397–422.

Baltes, M., Mayr, U., Borchelts, M., Maas, I., & Wilms, H. (1993). Everyday competence in old and very old age: An inter-disciplinary perspective. *Aging and Society*, *13*, 657–680.

Baltes, M. M., & Wahl, H. W. (1992). The dependency-support script in institutions: Generalization to community settings. *Psychology and Aging*, *7*, 409–418.

Berg, C. A., & Klaczynski, P. A. (1996). Practical Intelligence and Problem Solving: Searching for Perspectives. In F. Blanchard-Fields & T. M. Hess (Eds.), *Perspectives on cognitive change in adulthood and aging* (pp. 323–357). New York: McGraw-Hill.

Bieman-Copland, S., & Ryan, E. B. (1996). At my age my forgettery is better than my memory: An intergenerational analysis of age excuse behavior. Third International Conference on Communication, Aging and Health, Kansas City, KS.

Bieman-Copland, S., & Ryan, E. B. (in press). Age biased beliefs about memory successes and failures. *Journal of Gerontology: Psychological Sciences.*

Blanchard-Fields, F., & Hess, T. M. (1996). *Perspectives on cognitive change in adulthood and aging.* New York: McGraw-Hill.

Bosman, E. A. (1993). Age related differences in the motoric aspects of transcription typing skill. *Psychology and Aging, 8,* 87–102.

Botwinick, J. (1984). *Aging and behaviour* (3rd Ed.). New York: Springer.

Brigham, M. C., & Pressley, M. (1988). Cognitive monitoring and strategy choice in younger and older adults. *Psychology and Aging, 3,* 249–257.

Carstensen, L. L. (1991). Selectivity theory: Social Activity in life-span context. *Annual Review of Gerontology & Geriatrics, 11,* 195–213.

Cavanaugh, J. C. (1990). *Adult development and aging.* Belmont, CA: Wadsworth.

Cavanaugh, J. C., Grady, J. G., & Perlmutter, M. (1983). Forgetting and the use of memory aids in 20 to 70 year olds' everyday life. *International Journal of Aging and Human Development, 17,* 113–122.

Charness, N. (1981a). Aging and skilled problem solving. *Journal of Experimental Psychology: General, 110,* 21–38.

Charness, N. (1981b). Search in chess. Age and skill differences. *Journal of Experimental Psychology: Human Perception and Performance, 7,* 467–476.

Charness, N. (1985). Aging and problem-solving performance. In N. Charness (Ed.), *Aging and human performance.* Toronto: John Wiley & Sons.

Charness, N. (1995, August). Senior friendly input devices: Is the pen mightier than the mouse? Paper presented at the 103rd annual meeting of the American Psychological Association, New York, NY.

Charness, N., & Bosman, E. A. (1990). Human factors and design for older adults. In J. E. Birren & K. W. Schaie (Eds.), *Handbook of the psychology of aging* (3rd Ed., pp. 446–463). New York: Academic Press.

Charness, N., & Bosman, E. A. (1995). Compensation through environmental modification. In R. A. Dixon & L. Backman (Eds.), *Compensating for psychological deficits and declines: Managing losses and promoting gains.* Mahwah, NJ: Lawrence Erlbaum Associates.

Craik, F. I. M. (1977). Age differences in human memory. In J. E. Birren & K. W. Schaie (Eds.), *Handbook of the psychology of aging.* New York: Van Nostrand Reinhold.

Craik, F. I. M., Byrd, M., & Swanson, J. M. (1987). Patterns of memory loss in three elderly samples. *Psychology and Aging, 87,* 79–86.

Craik, F. I. M., & Salthouse, T. A. (1992). *The handbook of aging and cognition.* Hillsdale, NJ: Lawrence Erlbaum Associates.

Czaja, S., Hammond, K., Blascovich, J., & Swede, H. (1986). Learning to use a word processing system as a function of training strategy. *Behavior and Information Technology, 5,* 203–216.

DiPietro, L., Seeman, T. E., Merrill, S. S., & Berkman, L. F. (1996). Physical activity and measures of cognitive function in healthy older adults: The MacArthur Study of successful aging. *Journal of Aging and Physical Acitivity, 4,* 362–376.

Dittman-Kohli, F. (1990). The construction of meaning in old age: Possibilities and constraints. *Aging and Society, 10,* 279–294.

Dixon, R. A., & Backman, L. (1995). Concepts of compensation: Integrated, differentiated, and janus-faced. In R. A. Dixon & L. Backman (Eds.), *Compensating for psychological deficits and declines: Managing losses and promoting gains* (pp. 3–21). Mahwah, NJ: Lawrence Erlbaum Associates.

Elias, P. K., Elias, M. F., Robbins, M. A., & Gage, P. (1987). Acquisition of word-processing skills by younger, middle-aged, and older adults. *Psychology and Aging, 2,* 340–348.

Erber, J. T. (1989). Young and older adults' appraisal of memory failure in young and older target persons. *Journal of Gerontology: Psychological Sciences, 44,* 170–175.

Erber, J. T., & Rothberg, S. T. (1991). Here's looking at you: The relative effect of age and attractiveness on judgments about memory failures. *Journal of Gerontology: Psychological Sciences, 46,* 116–123.

Erber, J. T., Szuchman, L. T., & Rothberg, S. T. (1990a). Age, gender, and individual differences in memory failure appraisal. *Psychology and Aging, 5,* 600–603.

Erber, J. T., Szuchman, L. T., & Rothberg, S. T. (1990b). Everyday memory failure: Age differences in appraisal and attribution. *Psychology and Aging, 5,* 236–241.

Finch, C. E., & Schneider, E. L. (1985). *Handbook of the Biology of Aging* (2nd Ed.). New York: Van Nostrand Reinhold.

Fletcher, P. C., & Hirdes, J. P. (1996). A longitudinal study of physical activity and self-rated health in Canadians over 55 years of age. *Journal of Aging and Physical Activity, 4,* 136–150.

Heckhausen, J., & Lang, F. R. (1996). Social construction and old age: Normative conceptions and interpersonal processes. In G. R. Semin & K. Fiedler (Eds.), *Applied social psychology* (pp. 374–399). London: Sage.

Hiatt, L. G. (1987). Supportive design for people with memory impairments. In A. Kalicki (Ed.), *Confronting Alzheimer's disease.* Owings Mills, MD: National Health Publishing.

Hultsch, D., Hertzog, C., & Dixon, R. A. (1990). Ability correlates of memory performance in adulthood and aging. *Psychology and Aging, 5,* 356–368.

Hummert, M. L. (1990). Multiple stereotypes of elderly and young adults: A comparison of structure and evaluations. *Psychology and Aging, 5,* 183–193.

Kausler, D. H. (1994). *Learning and memory in normal aging.* San Diego, CA: Academic Press.

Kelley, C. L., & Charness, N. (1995). Issues in training older adults to use computers. *Behaviour and Information Technology, 14,* 107–120.

Kelley, C. L., Charness, N., Mottram, M., & Bosman, E. (April, 1994). The effects of cognitive aging and prior computer experience on learning to use a word processor. Poster session presented at the Fifth Cognitive Aging Conference, Atlanta, GA.

Kite, M. E., & Johnson, B. T. (1988). Attitudes toward older and younger adults: A meta-analysis. *Psychology and Aging, 3,* 233–244.

Kliegl, R., Smith, J., & Baltes, P. B. (1986). Testing the limits, expertise, and memory in adulthood and old age. In F. Klix & H. Hagendorf (Eds.), *Human memory and cognitive capabilities: Mechanisms and performances.* North Holland: Elsevier.

Kliegl, R., Smith, J., & Baltes, P. B. (1989). Testing-the-limits, and the study of adult age differences in cognitive plasticity of a mnemonic skill. *Developmental Psychology, 25,* 247–256.

Kliegl, R., Smith, J., & Baltes, P. B. (1990). On the locus of magnification of age differences during mnemonic training. *Developmental Psychology, 26,* 894–904.

Lachman, M. E., Weaver, S. L., Bandura, M., Elliot, E., & Lewkowicz, C. J. (1992). Improving memory and control beliefs through cognitive restructuring and self generated strategies. *Journals of Gerontology, 47,* 293–299.

Langer, E. J., & Rodin, J. (1976). The effects of choice and enhanced responsibility for the aged: A field experiment in an institutional setting. *Journal of Personality and Social Psychology, 34,* 191–198.

Lawton, M. P. (1982). Competence, environmental stress, and the adaptation of old people. In M. P. Lawton, P. G. Windley, & T. O. Byerts (Eds.), *Aging and the environment: Theoretical approaches* (pp. 33–59). New York: Springer.

Marsiske, M., Lang, F. R., Baltes, P. B., & Baltes, M. M. (1996). Selective optimization with compensation: Lifespan perspectives on successful human development. In R. A. Dixon & L. Backman (Eds.), *Compensating for psychological deficits and declines: Managing losses and promoting gains* (pp. 35–79). Mahwah, NJ: Lawrence Erlbaum Associates.

Meyer, B. J. F., & Rice, G. E. (1989). Prose processing in adulthood: The text, the reader and the task. In L. W. Poon, D. L. Rubin, & B. A. Wilson (Eds.), *Everyday cognition in adulthood and late life.* New York: Cambridge University Press.

Mihalko, S. L., & McAuley, E. (1996). Strength training effects on subjective well-being and physical function in the elderly. *Journal of Aging and Physical Activity, 4,* 56–68.

O'Connor, R. J., Aenchbacher, L. E., & Dishman, R. K. (1993). Physical activity and depression in the elderly. *Journal of Aging and Physical Activity, 1,* 34–58.

Palmore, E. B. (1990). *Ageism: Positive and negative.* New York: Springer.

Park, D. C., Smith, A. D., Lautenschlager, G., Earles, J. L., Frieske, D., Zwahr, M., & Gaines, C. L. (1996). Mediators of long-term memory performance across the life span. *Psychology and Aging, 11,* 621–637.

Parr, W. V., & Siegert, R. (1993). Adults' conceptions of everyday memory failures in others: Factors that mediate the effects of target age. *Psychology and Aging, 8,* 599–605.

Poon, L. W. (1985). Differences in human memory with aging: Natural causes and clinical implications. In J. E. Birren & K. W. Schaie (Eds.), *Handbook of the psychology of aging* (2nd ed.). New York, NY: Van Nostrand Reinhold.

Robertson-Tchabo, E. A., Hausman, C. P., & Arenberg, D. (1976). A classic mnemonic for older learners: A trip that works. *Educational Gerontology, 1,* 215–226.

Rodin, J., & Langer, E. J. (1977). Long-term effects of a control relevant intervention with the institutionalized aged. *Journal of Personality and Social Psychology, 35,* 897–902.

Rodin, J., & Langer, E. J. (1980). Aging labels: The decline of control and the fall of self-esteem. *Journal of Social Issues, 36,* 12–29.

Rowe, J., & Kahn, R. (1987). Human aging: Usual and successful. *Science, 237*, 143–149.

Ryan, E. B., Giles, H., Bartolucci, G., & Henwood, K. (1986). Psycholinguistic and social psychological components of communication by and with the elderly. *Language and Communication, 6*, 1–24.

Ryan, E. B., Hummert, M. L., & Boich, L. (1995). Communication predicaments of aging: Patronizing behavior toward older adults. *Journal of Language and Social Psychology, 13*, 144–166.

Ryan, E. B., Kwong See, S., Meneer, W. B., & Trovato, D. (1992). Age-based perceptions of language performance among younger and older adults. *Communication Research, 19*, 311–331.

Ryan, E. B., & Laurie, S. (1990). Evaluations of older and younger adult speakers: The influence of communication effectiveness and noise. *Psychology and Aging, 5*, 514–519.

Ryan, E. B., Meredith, S. D., MacLean, M. J., & Orange, J. B. (1995). Changing the way we talk with elders: Promoting health using the Communication Enhancement Model. *International Journal of Aging and Human Development, 41*, 87–105.

Ryan, E. B., Szechtman, B., & Bodkin, J. (1992). Attitudes toward younger and older adults learning to use computers. *Journal of Gerontology: Psychological Sciences, 47*, 96–101.

Ryff, C. D. (1989). In the eye of the beholder: Views of psychological well-being among middle-aged and older adults. *Psychology and Aging, 4*, 195–210.

Salthouse, T. A. (1984). Effects of age and skill in typing. *Journal of Experimental Psychology: General, 113*, 345–371.

Salthouse, T. A. (1987). The role of experience in cognitive aging. In K. W. Schaie (Ed.), *Annual Review of Gerontology and Geriatrics, 7*, 135–158. New York: Springer.

Salthouse, T. A. (1991). *Theoretical perspectives on cognitive aging.* Hillsdale, NJ: Erlbaum.

Salthouse, T. A., & Somberg, B. L. (1982). Skilled performance: Effects of adult age, and experience on elementary processes. *Journal of Experimental Psychology: General, 111*, 176–207.

Schaie, K. W., & Willis, S. L. (1986). Can decline in adult intellectual functioning be reversed? *Developmental Psychology, 22*, 223–232.

Schulz, R. (1976). Effects of control and predictability on the physical and psychological well-being of the institutionalized. *Journal of Personality and Social Psychology, 33*, 563–573.

Scogin, F., Strorandt, M., & Lott, L. (1985). Memory skills training, memory complaints and depression in older adults. *Journal of Gerontology, 40*, 562–568.

Seligman, M. E. P. (1975). *Helplessness: On depression, development and death.* San Francisco: W. H. Freeman and Company.

Searleman, A., & Herrmann, D. (1994). *Memory from a broader perspective.* New York: McGraw Hill.

Teitelman, J. L., & Priddy, J. M. (1988). From psychological theory to practice: Improving frail elders' quality of life through control-enhancing interventions. *The Journal of Applied Gerontology, 7*, 298–315.

Verhaeghen, P., Marcoen, A., & Goossens, L. (1992). Improving memory performance in the aged through mnemonic training: A meta-analytic study. *Psychology and Aging, 7*, 242-251.

Wallis, V. (1994). *Two old women.* New York: Harper Perennial.

West, R. L. (1989). Planning practical memory training for the aged. In L.W. Poon, D. C. Rubin, & B. A. Wilson (Eds.), *Everyday Cognition in Adulthood and Late Life* (pp. 573–597). New York: Cambridge University Press.

Willis, S. L. (1991). Cognition and everyday competence. In K. W. Schaie (Ed.), *Annual Review of Gerontology and Geriatrics, 11*, 80–109. New York: Springer.

Willis, S. L. (1996). Everyday cognitive competence in elderly persons: Conceptual issues and empirical findings. *The Gerontologist, 36*, 595–601.

Zandri, E., & Charness, N. (1989). Training older and younger adults to use software. Special Issue: Cognitive Aging: Issues in research and application. *Educational Gerontology, 15*, 615–631.

11

INTERVENTIONS TO IMPROVE COGNITIVE, EMOTIONAL, AND SOCIAL COMPETENCE IN LATE MATURITY

Dolores Pushkar and Tannis Arbuckle

Centre for Research in Human Development
Department of Psychology
Concordia University

The dramatic increase of longevity that has occurred in developed societies in this century will be a doubtful benefit unless the quality of life in those additional years is at a level acceptable to those who live them. In later life, there are increased disability rates, a higher incidence of chronic illness with comorbidity, negative cognitive changes and increased dependency upon family and institutional support (Gatz, 1995). Such changes are likely to result in declining physiological and psychological abilities in the older years.

Bieman-Copland, Ryan, and Cassano in another chapter in this volume discussed major theoretical approaches to the study of how competence is maintained despite these negative age-related changes. In this chapter, we will consider whether applied psychological research has demonstrated a capability to improve the quality of life in old age by increasing competent functioning. We will examine research in the three psychological dimensions of competence necessary for the maintenance of independent and satisfying daily life at all adult ages, including old age: cognitive, social and emotional abilities (Masterpasqua, 1989; Pushkar, Arbuckle, Conway, Chaikelson, & Maag, in press). Within each of these dimensions, we will briefly examine the interventions that have demonstrated effectiveness in improving the competent functioning of the elderly individual.

COGNITIVE COMPETENCE

The maintenance of cognitive competence is clearly an important aspect of successful aging. Yet laboratory studies provide a gloomy picture, with significant age-associated declines in performance being reported across all cognitive domains, leading deficit theorists to conclude that there is an underlying reduction in the efficiency of the aging brain resulting in a generalized slowing of cognitive processes (e.g., Salthouse, 1996). However, numerous critics have pointed out that adulthood is also a period of cognitive development

Improving Competence across the Lifespan, edited by Pushkar *et al.*
Plenum Press, New York, 1998.

leading to the maintenance and further improvement of intellectual abilities (e.g., Baltes, 1987; Chappell, 1996). Further, the design of much of the laboratory research has led to inflated estimates of the magnitude of age declines in basic processes of attention and memory. The usual experiment compares a group of young university students to a sample of community-dwelling elders, many of whom will not have been in a formal test-of-cognitive-abilities situation since they completed their own schooling 40 or more years earlier. The tasks are typically novel and frequently involve time constraints, limiting participants' ability to draw on their past experience and accentuating the negative effects of age-related slowing. Further, the inferior performance of the older participants is always attributed solely to their older age even though their older age means they belong to a different cohort with a different history (Baltes & Schaie, 1976). Williams and Klug (1996) found that cohort differences in longitudinal change in intelligence test performance were larger than age differences, showing the fallacy of concluding that aging is the unique cause of older adults' poorer performance in between-age-group comparisons. Finally, estimates of normative age declines may be inflated by the greater likelihood that older samples include individuals at a pre-diagnosable stage of a dementing disease.

Although true age differences in cognitive competence are likely to be smaller and begin later in life than laboratory research suggests, even longitudinal data suggest that by age 75 most cognitive functions have begun to decline (e.g., Schaie 1983). However, individuals of all ages have a certain amount of untapped resources (Staudinger, Marsiske, & Baltes, 1995), including untapped cognitive resources (Baltes, Kuhl, & Sowarka, 1992). By capitalizing on this reserve capacity, individuals may raise performance levels, offsetting some part of the age-related decline. In addition, factors other than age can contribute to declining performance of older individuals on cognitive tasks. The societal expectation that old age is associated with cognitive decline may reduce the individual's motivation to perform at top efficiency or, conversely, may heighten anxiety levels and interfere with peak performance. Further, the literature on age changes in physical abilities suggests that at least part of the loss in physical fitness with age is attributable to disuse. The same possibility has been raised for mental abilities. In general, individuals who maintain an active, engaged lifestyle appear to show better maintenance of memory and intelligence in old age (Arbuckle, Gold, & Andres, 1986; Arbuckle, Gold, Chaikelson, & Lapidus, 1994; Gold, Andres, Etezadi, Arbuckle, Schwartzman, & Chaikelson, 1995; Gribbin, Schaie, & Parham, 1980).

These suggestions that age declines are to some degree remediable has led to various types of intervention, both in normal aging and in early stages of dementia. Because memory complaints are frequent, even among healthy independently living elders, much of the remediation effort has been directed at improving memory although some has targetted other intellectual abilities. Another area of cognitive intervention has been directed at maximizing the cognitive performance of adults in early stages of dementia. In addition to attempts at direct intervention, some researchers have used indirect interventions, designed to enhance performance on cognitive tasks by changing expectations regarding age-related cognitive loss, by reducing anxiety level, or by increasing a sense of personal control over cognitive abilities.

COGNITIVE INTERVENTIONS IN THE HEALTHY AUTONOMOUS ELDERLY

Memory. Direct training of memory has been done in a variety of ways including training in use of imagery and mnemonic devices to remember specific items of information such as face-name associations or items on shopping lists, teaching of strategies for organizing or elaborating the to-be-remembered information so as to enhance its memorability, and providing subjects with manuals describing appropriate memory techniques for specific situations (e.g., Andrewes, Kinsella, & Murphy, 1996; Kotler-Cope & Camp, 1990). Studies have varied in terms of how the mnemonic aids have been taught, (e.g., by a trainer or a self-help manual), whether one or multiple techniques have been taught, and whether memory training has been supplemented by some other form of performance-enhancing manipulation.

Verhaeghen, Marcoen, and Goosens (1992) did a meta-analysis of 33 studies of memory training, all involving healthy community-dwelling elders. All studies met specified design criteria that included the use of some form of mnemonic training with pre- and post-training assessment of memory performance. Studies varied in terms of number of training conditions, in whether social contact or no contact control groups were used, and in whether generalization of training effects to untrained aspects of memory was tested. The meta-analysis showed larger pre- to post-training gains in training conditions than in control conditions, supporting the hypothesis that memory performance of older adults can be improved. All types of memory training were equally effective, being trained in more than one method was no more effective than being trained in only one, and there was little evidence of any generalization of training effects to untrained aspects of memory. Subsequent work (e.g., Andrewes et al., 1996; Hill, Allen, & McWhorter, 1991; Scogin & Prohaska, 1992; Stigsdotter Neely & Bäckman, 1993, 1995) has been largely consistent with the conclusions of the meta-analysis in that memory training of any kind is effective in enhancing memory performance of older adults but only on the type of memory task that was trained. Evidence on the durability of the effects is inconclusive. Camp, Foss, Stevens, Reichard, McKitrick, and O'Hanlon (1993), in their review of the memory intervention literature, concluded that any effects were of short duration. However, in a relatively large and well-controlled study, Stigsdotter Neely and Bäckman (1993, 1995) found that positive training effects were still apparent 3.5 years after the initial training.

Other Cognitive Abilities. Studies of the effects of training on other cognitive abilities have primarily examined measures of fluid intelligence, including figural relations, inductive reasoning, spatial abilities and response speed. The findings suggest that specific abilities can be improved through training, but generalization of training effects is limited (e.g., see review by Willis, 1989). For example, five hours of training on specific exemplars of either reasoning or spatial abilities tests were sufficient to reverse reliable declines on the trained dimension but did not improve performance on the untrained dimension (Schaie & Willis, 1986). Baltes, Sowarka and Kliegl (1989) found experimenter training and self-training methods were equally effective, but neither generalized to untrained skills. Willis (1989) argues that effects of training on particular tasks should be expected to generalize only to tasks that involve similar cognitive operations and strategies. For purposes of generalization, it may be more effective to train more general strategies than to focus training specifically on elements of the task. In line with this argument, Kramer, Larish and Strayer (1995) found that training participants to adopt a flexible strategy in allocating attention enhanced both learning and transfer for both young and old adults.

As in the case of memory training, training of other cognitive abilities may have limited durability. Willis and Nesselroade (1990) found a positive effect of cognitive training on figural relations after 5 years but Hayslip, Maloy and Kohl (1995) needed a reminder session to obtain a positive effect of training in inductive reasoning after 3 years and even then the effect was smaller than that of a non-cognitive intervention.

COGNITIVE COMPLAINTS

Complaints about problems in cognitive functioning, usually expressed in the form of concerns about declining memory, are relatively frequent among older adults. Consequently, several studies have examined the impact of memory training on subjective assessments of memory and cognitive functioning, and, in some cases, have compared the effectiveness of memory training for improving self-assessments of memory performance with its effectiveness for improving performance on objective measures of memory. In a recent meta-analysis of this literature, Floyd and Scogin (1997) found that while memory training did produce a significant improvement in post-training measures of subjective memory, the actual effect size (.19) was considerably smaller than the effect size reported by Verhaeghen et al. (1992) in their meta-analysis of the impact of memory training on objective measures of memory (.66). Further, when Floyd and Scogin (1997) compared the effect sizes of training on objective versus subjective measures of memory for the 10 studies that had used both outcome measures, this within-study comparison likewise showed smaller effects of training for subjectively assessed memory. Thus, while training enhances actual memory performance, it is less effective in changing people's concerns about their memory competence. Recent work in our own laboratory has shown that subjects' complaints about attentional problems are not related to objective measures of the ability to focus on relevant information and to inhibit irrelevant information, but rather are related to a personality characteristic of being hyperanxious (Béliveau, 1997). Watson and Pennebaker (1989) have shown that this personality characteristic, which they term negative affectivity, also inflates self-reported health complaints and Seidenberg, Taylor, and Haltiner (1994) have shown a similar association between negative affectivity and a variety of self-reported cognitive difficulties in adults aged 25 to 88 years. If memory complaints are in part a reflection of this personality characteristic, then it is perhaps not surprising that attempts to alleviate them through memory training have had only limited success. Bieman-Copland et al. (this volume) express a similar view, pointing out that the goal of complainants is to feel better about their memory, not necessarily to remember better.

COGNITIVE TRAINING IN DEMENTIA

In their 1990 review of the memory training literature, Kotler-Cope and Camp noted that relatively few interventions had been done with patients suffering from senile dementias. One reason for this paucity of research is that initial attempts to train dementia patients to use imagery mnemonics or to manage memory problems met with only limited success (e.g., Bäckman, Josephsson, Herlitz, & Stigsdotter, 1991; Hill, Evankovich, Sheikh, & Yesavage, 1987; Zarit, Zarit, & Reever, 1982). More recent work, however, has had some success using a variety of innovative techniques including active cognitive stimulation (Quayhagen, Quayhagen, Corbeil, Roth, & Rodgers, 1995), spaced retrieval training (McKitrick, Camp & Black, 1992), a significant event technique (Sandman, 1993) and memory wallets (Bourgeois & Mason, 1996). The success of these techniques appears to rest on one or more of the following elements: (a) they make use of external aids to memory rather than relying solely on internally generated ones, (b) they de-emphasize memorization as a goal and minimize the need to deliberately try to recall information, (c) they provide the patient with large amounts of positive feedback, and (d) they are specifically directed at maintaining information that is important in some way to the patient's well-being. For example, in the memory wallet technique used by Bourgeois and Mason (1996), patients in a day-care setting were given a wallet containing a set of pictures and sentences about people, places and events that were significant to them and that they had difficulty remembering. Provision of the wallet resulted in increased factual statements and decreased ambiguity and unintelligibility in patients' conversations with day-care workers.

LIMITS ON EFFECTIVENESS OF DIRECT COGNITIVE INTERVENTIONS

While the research shows that cognitive training can successfully enhance performance of older adults, the effects are limited in scope and possibly in durability. Further, the magnitudes of the training gains decrease with older age and with increasing cognitive impairment (Hill, Yesavage, Sheikh, & Friedman, 1989; Yesavage, Sheikh, Friedman, & Tanke, 1990). These findings are consistent with research using a testing-the-limits procedure to determine the actual amount of reserve capacity (Kliegl, Smith, & Baltes, 1989; 1990). This research has shown that reserve capacity for cognitive abilities decreases with age and with risk status for dementia (Baltes, et al. 1992; Baltes, Kuhl, Gutzmann, & Sowarka, 1995). In contrast to age and current cognitive ability, education does not seem to be a limiting factor, as even older adults with four years or less of education are able to benefit from training (Fernandez-Ballesteros & Calero, 1995).

INDIRECT INTERVENTIONS TO ENHANCE COGNITIVE PERFORMANCE

As previously noted, cognitive performance of older adults may be reduced by factors other than cognitive ability. Interventions to alleviate non-cognitive deterrents to efficient performance on cognitive measures have included relaxation training to reduce stress associated with the unfamiliar experimental situation or with concerns about possible cognitive declines (e.g., Hayslip, 1989; Yesavage, 1984) and cognitive restructuring to change expectations with respect to age-related changes in cognitive abilities (e.g., Caprio-Prevette & Fry, 1996; Lachman, Weaver, Bandura, Elliott, & Lewkowicz, 1992). Results have been generally positive in that, when used alone, such interventions tend to be at least as effective as direct cognitive training (e.g., Caprio-Prevette & Fry, 1996; Hayslip, 1989; Hayslip et al., 1995) and, when used in combination with memory training procedures, they tend to enhance the effectiveness of the latter on both objective and subjective measures of memory (Floyd & Scogin, 1997; Verhaegen et al., 1992).

CONCLUSIONS

Cognitive competence of both normally aging elders and those in the early stages of a dementing disease can be enhanced through interventions of various types. The effects of such interventions tend to be very specific. In the case of the dementing elderly, relatively modest interventions in terms of the time and effort needed to carry them out can make an appreciable improvement in the quality of life. For the healthy elderly, there has yet to be a clear demonstration of the long-term value of specific interventions. The finding that subjective memory assessments are not markedly changed by memory training also suggests caution in the interpretation of memory complaints as necessarily indicating some form of memory deficit. Cognitive functions are part of the whole person, and attempts to maintain either objective or subjective cognitive competence without taking into consideration other domains of competence may not be the most effective way of proceeding.

EMOTIONAL COMPETENCE

As Nancy Eisenberg illustrated in a previous chapter in this volume, the processes involved in emotional competence are varied and complex. But the essential requirement of functioning with emotional competence in daily life is emotional self-regulation to allow the experience of reasonable levels of well-being by avoiding depression and maintaining life satisfaction.

APPLIED RESEARCH WITH DEPRESSIVE STATES

There are difficulties in assessing depression in elderly samples, which appear partly due to differences in the way that depression is expressed at different ages (Kaszniak & Scogin, 1995). In elderly samples, depression may be expressed less through consciously experienced affective states and more through physiological complaints, resulting in a lower prevalence rate of diagnosed depressive disorder using DSM-IV criteria. The epidemiological data indicate that although older adults frequently report symptoms of depression, what has been called minor depression or dysthymia, the prevalence of diagnosed major depression is not high and indeed appears lower than the rates for younger people (George, 1993; Gatz, Kasl-Godley, & Karrel, 1996). Other data indicate that older adults identify fewer symptoms related to depression than do younger adults (Davies, Sieber & Hunt, 1994). This failure to recognize depressive symptoms as psychologically meaningful combined with a lower participation rate of depressed older people in epidemiological studies contributes to the lower rates of older people seeking help and being diagnosed as depressed, and to an underestimation of prevalence of depression among older adults (Thompson, Heller, & Rody, 1994).

Cohen (1990) argues that primary depression in elderly populations, that is depression in the absence of physical disorders or drug side effects, has a low incidence rate but since the elderly have higher frequencies of physical illness and medication usage, they are more at risk for secondary depressions resulting from these conditions than any other age group. The depletion syndrome has been postulated (Newman, Engel, & Jensen, 1991) as a depressive pattern, unlikely to be diagnosed by standard criteria, peculiar to elderly people, characterized by feelings of worthlessness, lack of interest, loss of appetite, and thoughts of death and dying. Although such symptoms may not result in standard diagnostic classifications, the depressive symptoms can impair functioning in individuals who could benefit from treatment. Gatz et al. (1996) concluded that although clinical depression does not increase with age, dysphoria and sub-clinical symptoms of depression do. It is possible that the inability to find meaning in the present life context leading to feelings of alienation, plays a precipitating role.

Although therapists lacking experience with elderly clients may be more skeptical, those who work with the elderly are more convinced of psychotherapeutic effectiveness with geriatric samples (Knight, 1993). The effectiveness of psychosocial interventions has been recently examined in a meta-analysis of 17 recent studies (Scogin & McElreath,

1994). The results of the analysis revealed that psychosocial interventions, including cognitive, behavioral, reminiscence, psychodynamic and eclectic approaches, were significantly more effective for both major depression and subclinical depression than no-treatment and placebo control groups. The median effect size for psychosocial interventions compared to the control conditions was .78, comparable to those found with depressed adults of all ages. The significant effects for psychosocial interventions were maintained when both self-report and clinician ratings were used as outcome measures and for individual and group therapy. Sufficient data were available only to allow the examination of specific effect sizes for cognitive and reminiscence therapies and both were found to be more effective compared to control groups.

Other reviews of the effectiveness of psychotherapeutic interventions with elderly clients have reached similar conclusions (Gallagher-Thompson & Thompson, 1995a, 1995b; Niederehe, 1994; Reynolds, Frank, Houck, Mazumdar, Dew, Cornes, Buysse, Begley, & Kupfer, 1997; Terri, Curtis, Gallagher-Thompson, & Thompson, 1994). Gallagher-Thompson and Thompson (1994) point out that modifying treatment may overcome problems of some elderly clients, such as sensory problems and reduced memory capacity. For example, a period of socialization for elderly clients so that expectations of client and therapist behaviour can be made explicit and incorrect beliefs can be elicited facilitates the process of psychotherapy with elderly clients.

Research findings suggest different success rates for interventions targeting specific factors associated with depression. Jorm (1995) argues that interventions designed for depressed elderly people, where situational factors play a prominent predisposing role such as physical illness or acting as a caregiver for family members, are likely to be effective in reducing depression. In contrast, elderly people with predisposing personal conditions, such as prior history of depression and personality traits predisposed to depressive reactions, are particularly vulnerable to depression in late life and are less likely to benefit from interventions. Similarily, programmes for at-risk populations, such as recently disabled and recently bereaved older people, have reduced psychological distress (see for example Reich & Zautra, 1989), and for nursing home residents (Dhooper, Green, Huff, & Austin-Murphy, 1993).

A variety of research findings provide a basis for the further development of interventions to enhance affect self-regulation in elderly people. Contextual effects have been examined. For example, Lawton, DeVoe, and Parmelee (1995) examined the influence of environmental context on mood, using an adaptation-level theoretical framework. They found that the mood of elderly residents in geriatric centres was significantly related to the quality of events they experienced in daily institutional life, regardless of level of psychopathology. Further research determining the type of events that are experienced as novel, challenging, security inducing and pleasureful for older residents could provide a basis for institution-based interventions.

Other research has provided information about psychological processes in older people with higher rates of depressive symptoms, which can provide a basis for the development of therapeutic interventions. Thompson and Heller (1993), for example, found that older depressed women, especially those leading lives with high levels of daily routines, were most likely to have lower quantitative and qualitative inter-personal problem solving abilities. Interpersonal conflict was especially difficult for depressed women to handle. This finding suggests a social skill remediation program could be useful for such individuals.

Basic research theories and data can also provide useful guides to therapeutic interventions for depressed elderly people. Allard and Mishara (1995), guided by Petrie's 1967 research on personality-perceptual styles of modulating stimulus intensity, found that perceptual augmentation was related to depression without manic features and perceptual reduction was related to depression with manic symptoms in elderly community-dwelling adults. Hypothesizing that sensory deprivation, caused by factors such as chronic illness, sensory disabilities, loss of meaningful and stimulating activities, influences the development of depression for older people who are inclined to dampen the effects of incoming stimulation, Allard and Mishara argue that depressed elderly people with manic symptoms would benefit from interventions that emphasize complexity, variability and stimulating activities. In contrast, depressed elderly people without manic symptoms, who are inclined to magnify the impact of stimulation, would benefit more from interventions which emphasize predictability, internal structure and avoid strong and variable stimulation.

Other research has examined various methods of delivery of interventions, for example, home visits for clients with physical mobility problems, which are common among elderly samples. The Institute of Medicine report on Prevention of Mental Disorders (Munôz, Mrazak, & Haggerty, 1996) concluded that the provision of preventive services for a broad range of problems in non-mental health settings would increase opportunities for treatment. Hanser and Thompson (1994), for example, used home-visits to test the efficacy of music therapy for homebound depressed elderly clients. Other research has targeted rural elderly, for whom distance and transportation issues are important considerations (Korte, 1990). Targeting the therapist rather than the client has also been examined with less benefit being demonstrated. For example, the provision of informational support to primary care physicians increased the number of diagnoses of depression and prescription of anti-depressants but had no effect on use of drugs that cause depression, rate of psychiatric referrals or patient outcomes (Callahan, Hendrie, Dittus, Brater, Hui, & Tierney, 1994).

CONCLUSIONS

Although a complete review of the research examining the effects of interventions in alleviating depression in elderly people is beyond the scope of this chapter, it appears that depression in elderly people can be as successfully treated using psychotherapeutic techniques as in other life stages. At all life stages, depression associated with particular precipitating situations, in contrast to personality-linked vulnerability, is more amenable to interventions. Age-related characteristics and special needs of the various elderly populations necessitate particular attention and may require modification both in therapeutic techniques and delivery of services.

WELL-BEING

The other important task of emotional self-regulation is the maintenance of reasonable levels of well-being. Longitudinal research has demonstrated great stability in individual levels of well-being across the adult lifespan and no apparent decrease in subjective well-being in later life (Argyle & Martin, 1991; Brandtstädter & Wentura, 1995; Costa, Zonderman, McCrae, Cornoni-Huntley, Locke, & Barbano, 1987). Abundant research has examined the determinants of well-being and has found that both objective life circumstances and personality characteristics determine well-being. Health, socioeconomic variables, integration in a satisfactory social network, as assessed by both objective and subjective methods, predict well-being (see for example Antonucci, 1985; Argyle & Martin, 1991; Chappell, 1995; Diener, 1984; Rubinstein, Lubben, & Mintzer, 1994). Basic personality traits appear to be more important determinants of well-being than objective life situation factors (Costa & McCrae, 1980; George, 1985). Considering the durability and complexity of the other determinants of well-being, it is not surprising that most interventions targeting well-being have focussed on social functioning.

Theoretical contributions, in addition to the work of Baltes and Baltes (1990), have provided frameworks for development of interventions targeting the social functioning of elderly people in attempts to improve levels of well-being. Atchley (1989, 1993) emphasizes the importance of the older person's being able to maintain continuity of important central roles, activities and inter-personal relations that have functioned well in earlier adult stages, especially those that have helped to affirm the constancy and value of individual experience of the self. Similarly, role theory has demonstrated that being engaged voluntarily in a number of high quality roles (Moen, 1995) benefits well-being.

In the social domain, Antonnucci (1985) has developed the convoy model emphasizing the effects of lifespan social support on well-being. This theory states that individuals travel through life in a social convoy composed of family members and friends, who provide emotional, social and instrumental support. Although the composition of the outer layers of the convoy may change with particular life contexts and needs, the existence of a satisfactory core at the social convoy is essential for the well-being of most individuals.

Socioemotional selectivity theory examines the processes of selecting out and retaining members in the social network, focussing on older individuals' proactive role in constructing their own social lives (Carstensen, Hanson, & Freund, 1995; Lang & Carstensen, 1994). The theory assumes that emotional regulation is of primary importance and social interactions are selected in accordance with their success in maintaining emotional benefits. The reduction in social interaction that occurs in late life is partly a function of individuals selecting out peripheral relationships while emphasizing and retaining close emotional relations with family members and friends who facilitate the individual's well-being.

These theories converge in suggesting that interventions designed to enhance well-being employing social roles, activities and support will be more successful if older individuals are recognized as playing a central role in selecting and rejecting aspects of interventions that meet their needs. Further, it suggests that interventions that help to establish or re-establish activities and interpersonal relations based on the same values that maintained past patterns will be most successful. Increases in social contacts and activities without due consideration of the match between the characteristics of the individual and the intervention are unlikely to be successful. In practical terms, this suggests that recognition of the heterogeneity of older people be built into intervention programmes and that individual choice and control is essential for success.

A meta-analysis of 31 studies examined the effects of interventions enhancing control, psychoeducational interventions, and social activities on the well-being of institutionalized and community-based older subjects (Okun, Olding, & Cohn, 1990). A median post-treatment effect size of .67 was found with the three conditions achieving immediate post-treatment significant effect sizes, but not differing significantly among each other. However, the effect sizes became insignificant over time. The results of this meta-analysis indicate that improving well-being for any appreciable length of time is a difficult goal to attain, presumably because well-being is influenced by factors that are difficult to change, such as long-standing personality traits (Costa et al., 1987; Costa & McCrae, 1994) and objective life situation factors.

SOCIAL ROLES AND ACTIVITIES

Volunteering is a valuable role option for many older people, allowing older people to maintain or re-establish reciprocal helping relationships, and providing continuity with earlier established independent and competent adult lifestyles. Although there are some inconsistent findings, volunteers have been found in many studies to have higher levels of well-being and health (Chambre, 1988). However, a causal relationship between well-being and volunteering has not been well established and longitudinal, controlled intervention research in this area is needed (Fisher & Schaffer, 1993). It is likely that more isolated older people who lack social resources would gain particular benefits from volunteering.

Many programmes have successfully increased social activities among community and institutionally-based older people. However, Jerrome (1991) points out that associations and programs organized by older people for themselves differ from those organized by others. Those run by older people appear to be more sensitive to their needs for intimacy and feelings of effectiveness and pay more attention to the social processes of friendship and leadership formation within the group. Even in programmes that have non-social objectives, social motivations have an important influence on participation. Duncan, Travis, and McAuley (1995), for example, found that social participation and sense of belonging to a community were important reasons for becoming involved in a physical activity programme. Due consideration must be given to the level of functioning of participants and the degree of initiative that staff must display to involve residents with differing degrees of abilities and impairment, especially in institutional-based populations (WindRiver, 1993).

SOCIAL INTEGRATION

Programmes employing visitors trained to increase the building of networks for isolated elderly have demonstrated some success (Korte & Gupta, 1991). Some programmes have also failed to improve well-being through increasing contact (Heller, Thompson, Trueba, Hogg, & Vlachos-Weber, 1991; Heller, Thompson, Vlachos-Weber, Steffen, & Trueba, 1991). Heller and colleagues argue that interventions should emphasize reinforcing meaningful role activities and strengthening existing ties in the life context of the older person before trying to create new ones. Others have emphasized the importance of intimacy in interpersonal exchanges (Barrera, 1991) and pointed out that some lost meaningful ties may not be readily or ever replaced by other relationships (Gatz, 1995). These recommendations are in accordance with the theoretical frameworks regarding the social functioning of older individuals outlined above.

Interventions have targetted the role of social support networks in helping people safeguard well-being when confronted with major illness or medical procedures (Oxman & Hull, 1997), with generally positive results. Social support is particularly important for elderly individuals with increased rates of disabilities. Typically, individuals who experience declining function also experience reduced social contacts (Newsom & Schulz, 1996), due to such factors as activity restrictions, resulting inability to maintain reciprocity in interactions, feelings of burden, which are associated with lower feelings of well-being and increased risk of depression. Support from friends appears to be more critical in maintaining well-being than support from family, although family support becomes more important with increasing disability (Dean, Kolody, & Wood, 1990; Cicerelli, 1981). These findings suggest that targetting the maintenance of social ties outside of family members would be a useful approach to maintaining well-being for elderly people with disabilities.

One of the most striking developments in this area is the increased number of self-help groups of elderly people, presumably based on the assumption that "it takes one to help one" (Pillemer & Suitor, 1996). Self-help group interventions, for example, have been found to be effective in helping people to cope with life transitions such as bereavement (Caserta & Lund, 1993), especially for people with lower levels of resources and competencies. It is possible, however, the group component might not be necessary to achieve benefits in well-being. Pillemer and Suitor (1996) found that among family caregivers of relatives with dementia, caregivers who had more members in their social support network with similar experiences had significantly less depression. They point out that relying on group interventions is not necessary and that mobilizing caregivers with similar experiences in the existing network or arranging one-on-one support would also be effective.

CONCLUSIONS

This brief review should indicate that attempts to improve well-being in elderly people, although achieving some success in helping people cope with life stressors, have not demonstrated considerable long-term success. As expected from the theoretical frameworks summarized above, it is difficult to create interventions to replace or create the long term meaningful, intimate social integration that is central to most people's well-being. Some data suggest that it might be easier to change people's perceptions of the situation than to effect objective change (see Brand, Lakey & Berman, 1995 below). This suggestion is in agreement with previous findings that cognitive reframing is one of the most effective and frequently used ways of coping employed by older adults (see for example, Diehl, Coyle, & LaBouvie-Vief, 1996).

It appears to be easier to involve people in group activities and roles that are satisfying, but the translation of such involvement into causal effects on well-being has yet to be demonstrated. It is quite possible that people who are willing to engage in such activities are self-selected on the basis of, among other characteristics, higher levels of well-being and more effective coping skills.

SOCIAL COMPETENCE

Competence in interpersonal or social functioning has received less attention in research with elderly people than among younger populations. As indicated above, some interventions designed to alleviate depression and to increase well-being have targetted the improvement of interpersonal problem solving skills or increasing or remobilizing the social support networks of elderly people. Since social competence and extent and quality of social support network are associated (Chappell, 1995), an overlap in the nature of interventions designed to improve emotional and social competence is to be expected.

Some social skills training programmes have been developed for independently living community-based older people. For example, Brand, Lakey and Berman (1995) trained single adults in social skills and cognitive reframing techniques regarding the self and relations with others. The programme was successful in increasing the perceived level of support from family members, however, there was more change in cognitions about the self than about perception of support. These results parallel the results of attempts to change expectations in the cognitive domain discussed above.

But most social skills training programmes for older people have been developed primarily for populations with disabilities (Donahue, Acierno, Hersen, & Van Hasselt, 1995), elderly people in residential care (Fisher & Carstensen, 1990), and people in early stages of dementia who display aggressive and inappropriate behaviour (Doyle, 1993; Vaccaro, 1992). Similarly, some programmes have been developed to improve the communication skills of elderly people who have suffered impairments through age-related disease processes (Bourgeois, 1991; Jordan, Worrall, Hickson, & Dodd, 1993). Activity programmes have also been designed for long-term residents in care facilities to reduce or delay functional impairment in such areas as social behaviour and communication (Sulman & Wilkinson, 1989). Generally, these programmes have achieved some success when targeting particular behavioral patterns.

Programmes have also successfully trained staff in techniques to improve the social functioning and quality of life of residents. Based upon their previous research identifying behavioral systems in the natural ecology of long-term care institutions, Baltes, Neumann, and Zank (1994) were able to implement a training programme for staff in communication skills, knowledge about aging and principles of behavioral modification. The programme significantly decreased dependence-supportive staff behaviour and resulted in significantly increasing independent behaviour of residents. On the other hand, some programmes designed to improve the quality of life of elderly residents by targeting the knowledge and performance of staff have also failed to demonstrate benefits either in staff performance or residents' outcome (Smyer, Brannon, & Cohn, 1992). Smyer (1995) argues that failure to adequately understand the ramifications of organizational culture and structure help to explain the lack of effects of interventions. For example, obtaining the support of supervisors in the implementation of a skills training programme for their staff is necessary.

CONCLUSIONS

The brief review of studies in this area indicates some success in improving the social functioning of both institution and community-based older people, although modifying expectations may be as important as modifying behaviour. Community-based elderly people are probably less likely to participate in social skills programs while individuals with disabilities are more likely to be recognized by themselves or others as appropriate targets for such interventions. The more tightly controlled context of institutions also appears to make it easier to demonstrate the effectiveness of interventions than the community-based programmes in more open-ended contexts.

SOME CONSIDERATIONS FOR THE DEVELOPMENT OF INTERVENTIONS FOR ELDERLY SAMPLES

Shortage of space precludes a complete review of all relevant issues. An excellent discussion of the broad issues involved in the development and application of intervention and prevention programmes for elderly populations are found in Grams and Albee (1995). We will briefly discuss some issues that we believe have not been sufficiently emphasized, particularly for those readers who do not specialize in aging research.

Many of the problems of old age are not intrinsically due to the nature of aging. Many declines in function, commonly regarded as due to old age, are the result of other factors correlated with age which cause loss of function or acceleration of decline caused by intrinsic aging processes. Such factors include illness, trauma, long-term consequences of lifestyle choices, disuse effects, and role changes which remove external constraints, requirements and challenges which enhance functioning. When these effects are considered, it is reasonable to assume that the current level of functioning of the majority of elderly people is not near maximum potential. The research reviewed in this chapter demonstrates that the performance of older people can be somewhat enhanced, but clearly there remains room for increased efficacy of interventions. Consequently, it becomes extremely important to accurately differentiate the causes of decreased competence in old age and to determine which processes are most amenable to interventions. In addition, considering the situational-specific nature of competencies, ideally the assessment of the functioning of the elderly individual should be done *in situ*. Finally, it is important to define the objective of an intervention; for example, to prevent the development of a problem, to change the individual's expectancies, to alleviate the problem by recovering function, or to compensate for problem functioning by developing alternative approaches (Salthouse, 1995).

It is also important to recognize the difference between individual abilities and competence in functioning in everyday life. Most elderly individuals reside in and function with environments that, to a large extent, are self-selected and created, with many built-in environmental and social supports that enable them to function reasonably adequately in their everyday activities. These supports allow the elderly person to lead relatively independent lives at a level much higher than would be predicted on knowledge of the individual's specific abilities alone. Any intervention will manipulate usually only one or two of the multitude of factors influencing the functioning of and outcomes for elderly clients. Undoubtedly, there are some factors influencing outcomes that are completely unknown to the practitioner. It becomes important to anticipate as much as possible the unintended side effects of interventions, which may well have negative effects on functioning, either in the target area or in an apparently unrelated functional area, or on the people in the environment who are providing support and enabling the elderly person to retain adequate degrees of autonomy.

Most interventions are designed to work primarily with individuals or families. However, as Gatz (1995) rightly points out, interventions at the community and policy levels influence many more people. Considering the importance of health and income as determinants of competence for older adults, such policy interventions become vital. One of the most important roles of researchers and practitioners in this area is that of educating officials and the public about aging.

Finally, as Grams and Albee (1995) emphasize, since lifestyle choices often have exceedingly long-term and long lasting effects in late life, in large part our experience of aging is determined by the events and nature of our earlier life. Thus, development across the lifespan must be managed to retain competence in old age. Ideally interventions should occur well before the older years so that the potential for productive and successful aging can be more fully realized.

REFERENCES

Allard, C., & Mishara, B. L. (1995). Individual differences in stimulus intensity modulation and its relationship to two styles of depression in older adults. *Psychology and Aging, 10*, 395–403.

Andrewes, D. G., Kinsella, G., & Murphy, M. (1996). Using a memory handbook to improve everyday memory in community-dwelling older adults with memory complaints. *Experimental Aging Research, 22*, 305–322.

Antonucci, T. C. (1985). Personal characteristics, social support, and social behaviour. In R. H. Binstock & E. Shanas (Eds.), *Handbook of aging and the social sciences* (2nd ed., pp. 94–128). New York: Van Nostrand Reinhold.

Arbuckle, T. Y., Gold, D., & Andres, D. (1986). Cognitive functioning of older people in relation to social and personality variables. *Psychology and Aging, 1,* 55–62.

Arbuckle, T. Y., Gold, D. P., Chaikelson, J. S., & Lapidus, S. (1994). Measurement of activity in the elderly: The Activities Checklist. *Canadian Journal on Aging, 13,* 550–565.

Argyle, M., & Martin, M. (1991). The psychological causes of happiness. In F. Strack, M. Argyle, & N. Schwarz (Eds.), *Subjective Well-Being.* Oxford: Pergamon Press.

Atchley, R. C. (1989). A continuity theory of normal aging. *The Gerontologist, 29,* 183–190.

Atchley, R. C. (1993). Continuity theory and the evolution of activity in later adulthood. In J. R. Kelly (Ed.), *Activity and aging: Staying involved in later life.* (pp 5–16). Newbury Park, California: Sage Publications.

Bäckman, L., Josephsson, S., Herlitz, A., & Stigsdotter, A. (1991). The generalizability of training gains in dementia: Effects of an imagery-based mnemonic on face-name retention duration. *Psychology and Aging, 6,* 489–492.

Baltes, P. B. (1987). Theoretical propositions of life-span developmental psychology: On the dynamics between growth and decline. *Developmental Psychology, 23,* 611–626.

Baltes, P. B., & Baltes, M. M. (1990). *Successful aging: Perspectives from the behavioral sciences.* New York: Cambridge University Press.

Baltes, M. M., Kuhl, K.-P., Gutzmann, H., & Sowarka, D. (1995). Potential of cognitive plasticity as a diagnostic instrument: A cross-validation and extension. *Psychology and Aging, 10,* 167–172.

Baltes, M. M., Kuhl, K.-P., & Sowarka, D. (1992). Testing for limits of cognitive reserve capacity: A promising strategy for early diagnosis of dementia? *Journal of Gerontology: Psychological Sciences, 47,* 165–167.

Baltes, M. M., Neumann, E. M., & Zank, S. (1994). Maintenance and rehabilitation of independence in old age: An intervention program for staff. *Psychology and Aging, 9,* 179–188.

Baltes, P. B., & Schaie, K. W. (1976). On the plasticity of intelligence in adulthood and old age: Where Horn and Donaldson fail. *American Psychologist, 31,* 720–725.

Baltes, P. B., Sowarka, D., & Kliegl, R. (1989). Cognitive training research on fluid intelligence in old age. What can older adults achieve by themselves? *Psychology and Aging, 4,* 217–221.

Barrera, M. (1991). Social support interventions and the third law of ecology. *American Journal of Community Psychology, 19,* 133–138.

Béliveau, M.-J. (1997). Attention, control, and off-target speech in older adults. Unpublished B.A. Honours thesis, Department of Psychology, Concordia University, Montreal, Quebec, Canada.

Bourgeois, M. (1991). Communication treatment for adults with dementia. *Journal of Speech and Hearing Research, 34,* 831–844.

Bourgeois, M. S., & Mason, L. A. (1996). Memory wallet intervention in an adult day-care setting. *Behavioral Interventions, 11,* 3–18.

Brand, E. F., Lakey, B., & Berman, S. (1995). A preventive, psychoeducational approach to increase perceived social support. *American Journal of Community Psychology, 23,* 117–135.

Brandtstädter, J., & Wentura, D. (1995). Adjustment to shifting possibility frontiers in later life: Complementary adaptive modes. In R. A. Dixon & L. Backman (Eds.), *Compensating for Psychological Deficits and Declines* (pp. 83–106). Mahwah, NJ: Lawrence Erlbaum Associates.

Callahan, C. M., Hendrie, H. C., Dittus, R. S., Brater, D. C., Hui, S. L., & Tierney, W. M. (1994). Improving treatment of late life depression in primary care: A randomized clinical trial. *Journal of the American Geriatrics Society, 42,* 839–846.

Camp, C. J., Foss, J. W., Stevens, A. B., Reichard, C. C., McKitrick, L. A., & O'Hanlon, A. M. (1993). Memory training in normal and demented elderly populations: The E–I–E–I–O model. *Experimental Aging Research, 19,* 277–290.

Caprio-Prevette, M. D., & Fry, P. S. (1996). Memory enhancement program for community based older adults: Development and evaluation. *Experimental Aging Research, 22,* 281–303.

Carstensen, L. L., Hanson, K. A., & Freund, A. M. (1995). Selection and compensation in adulthood. In R. A. Dixon & L. Backman (Eds.), *Compensating for Psychological Deficits and Declines* (pp. 107–126). Mahwah, NJ: Lawrence Erlbaum Associates.

Caserta, M. S., & Lund, D. A. (1993). Intrapersonal resources and the effectiveness of self-help groups for bereaved older adults. *Gerontologist, 33,* 619–629.

Chambre, S. M. (1988). *Good deeds in old age: Volunteering by the new leisure class.* Lexington, MA: Lexington Books.

Chappell, M. S. (1996). Changing perspectives on aging and intelligence: An empirical update. *Journal of Adult Development, 3,* 233–239.

Chappell, N. L. (1995). Informal social support. In L. A. Bond, S. J. Cutler & A. Grams (Eds.), *Promoting successful and productive aging* (pp. 171–185). Thousand Oaks, CA: Sage.

Cicerelli, V. G. (1981). *Helping elderly parents: The role of adult children.* Boston: Auburn House.

Cohen, G. D. (1990). Psychopathology and mental health in the mature and elderly adult. In J. E. Birren & K. W. Schaie (Eds.), *Handbook of the psychology of aging* (pp. 359–368). New York: Academic Press.

Costa, P. T., Jr., & McCrae, R. R. (1980). Influence of extraversion and neuroticism on subjective well-being: Happy and unhappy people. *Journal of Personality and Social Psychology, 38,* 668–678.

Costa, P. T., Jr., & McCrae, R. R. (1994). Set like plaster? Evidence for the stability of adult personality. In T. F. Heatherton & J. L. Weinberger (Eds), *Can Personality Change?* Washington, DC: American Psychological Association.

Costa, P. T., Jr., Zonderman, A. B., McCrae, R. R., Cornoni-Huntley, J., Locke, B. Z., & Barbono, H. E. (1987). Longitudinal analyses of psychological well-being in a national sample: Stability of mean levels. *Journal of Gerontology, 42,* 50–55.

Davies, R. M., Sieber, K. O., & Hunt, S. L. (1994). Age-cohort differences in treating symptoms of mental illness: A process approach. *Psychology and Aging, 9,* 446–453.

Dean, A., Kolody, B., & Wood, P. (1990). Effects of social support from various sources on depression in elderly persons. *Journal of Health and Social Behaviour, 31,* 148–161.

Diehl, M., Coyle, N., & Labouvie-Vief, G. (1996). Age and sex differences in strategies of coping and defense across the life span. *Psychology and Aging, 11,* 127–139.

Diener, E. (1984). Subjective well-being. *Psychological Bulletin, 95,* 542–575.

Dhooper, S. S., Green, S. M., Huff, M. B., & Austin-Murphy, J. (1993). Efficacy of a group approach to reducing depression in nursing home elderly residents. *Journal of Gerontological Social Work, 20,* 87–100.

Donahue, B., Acierno, R., Hersen, M., & Van Hasselt, V. B. (1995). Social skills training for depressed, visually impaired older adults: A treatment manual. *Behaviour Modification, 19,* 379–424.

Doyle, C. (1993). Social interventions to manage mental disorders of the elderly in long-term care. *Australian Psychologist, 28,* 25–30.

Duncan, H. H., Travis, S. S., & McAuley, W. J. (1995). An emergent theoretical model for interventions encouraging physical activity (mall walking) among older adults. *Journal of Applied Gerontology, 14,* 64–77.

Fernandez-Ballesteros, R., & Calero, M. D. (1995). Training effects on intelligence of older persons. *Archives of Gerontology and Geriatrics, 20,* 135–148.

Fisher, J. E., & Carstensen, L. L. (1990). Generalized effects of skills training among older adults. *Clinical Gerontologist, 9,* 91–107.

Fisher, L. R., & Schaffer, K. B. (1993). *Older volunteers: A guide to research and practice.* California: Sage Publications.

Floyd, M., & Scogin, F. (1997). Effects of memory training on the subjective memory functioning and mental health of older adults. *Psychology and Aging, 12,* 150–161.

Gallagher-Thompson, D., & Thompson, L. W. (1995a). Efficacy of psychotherapeutic interventions with older adults. *The Clinical Psychologist, 48,* 24–30.

Gallagher-Thompson, D., & Thompson, L. W. (1995b). Psychotherapy with older adults in theory and practice. In B. Bongar & L. Bentler (Eds.), *Comprehensive textbook of psychotherapy* (pp. 359–379). New York: Oxford.

Gatz, M. (1995). Questions that aging puts to preventionists. In L. A. Bond, S. J. Cutler, & A. Grams (Eds.), *Promoting successful and productive aging* (pp. 36–50). Thousand Oaks, California: Sage.

Gatz, M., Kasl-Godley, J. E., & Karel, M. J. (1996). Aging and mental disorders. In J. E. Birren & K. W. Schaie (Eds.), *Handbook of the psychology of aging* (pp. 365–377). New York: Academic Press.

George, L. K. (1985). The impact of personality and social status upon activity and psychological well-being. In E. Palmore, E. W. Busse, G. L. Maddox, J. B. Nowlin, & I. C. Siegler (Eds.) *Normal Aging III: Reports from the Duke Longitudinal Studies, 1975–1985.* Durham, NC: Duke University Press.

George, L. K. (1993). Depressive disorders and symptoms in later life. In M. A. Smyer (Ed.), *Mental Health and Aging: Progress and Prospects* (pp. 64–74). New York: Springer.

Gold, D. Pushkar, Andres, D., Etezadi, J., Arbuckle, T., Schwartzman, A., & Chaikelson, J. (1995). Structural equation model of intellectual change and continuity and predictors of intelligence in elderly men. *Psychology and Aging, 10,* 294–303.

Grams, A., & Albee, G. W. (1995). Primary prevention in the service of aging. In L. A. Bond, S. J. Cutler, & A. Grams (Eds.), *Promoting successful and productive aging* (pp. 5–35). Thousand Oaks, CA: Sage.

Gribbin, K., Schaie, K. W., & Parham, I. A. (1980). Complexity of lifestyle and maintenance of intellectual abilities. *Journal of Social Issues, 36,* 47–61.

Hanser, S. B., & Thompson, L. W. (1994). Effects of a music therapy strategy on depressed older adults. *Journals of Gerontology, 49,* 265–269.

Hayslip, B, Jr. (1989). Alternative mechanisms for improvements in fluid ability performance among older adults. *Psychology and Aging, 4,* 122–124.

Hayslip, B, Jr., Maloy, R. M., & Kohl, R. (1995). Long-term efficacy of fluid ability interventions with older adults. *Journal of Gerontology: Psychological Sciences, 50B,* 134–140.

Heller, K., Thompson, M. G., Trueba, P. E., Hogg, J. R., & Vlachos-Weber, I. (1991). Peer support telephone dyads for elderly women: Was this the wrong intervention? *American Journal of Community Psychology*, *19*, 53–74.

Heller, K., Thompson, M. G., Vlachos-Weber, I., Steffen, A. M., & Trueba, P. E. (1991). Support interventions for older adults: Confidante relationships, perceived family support, and meaningful role activity. *American Journal of Community Psychology*, *19*, 139–146.

Hill, R. D., Allen, C., & McWhorter, P. (1991). Stories as a mnemonic aid for older learners. *Psychology and Aging*, *6*, 484–486.

Hill, R. D., Evankovich, K. D., Sheikh, J. I., & Yesavage, J. A. (1987). Imagery mnemonic training in a patient with primary degenerative dementia. *Psychology and Aging*, *2*, 204–205.

Hill, R. D., Yesavage, J. A., Sheikh, J. I., & Friedman, L. (1989). Mental status as a predictor of response to memory training in older adults. *Educational Gerontology*, *15*, 633–639.

Jerrome, D. (1991). Loneliness: Possibilities for intervention. *Journal of Aging Studies*, *15*, 195–208.

Jordan, F. M, Worrall, L. E., Hickson, L. M., & Dodd, B. J. (1993). The evaluation of intervention programmes for communicatively impaired elderly people. *European Journal of Disorders of Communication*, *28*, 63–85.

Jorm, A. F. (1995). The epidemiology of depressive states in the elderly: Implications for recognition, intervention and prevention. *Social Psychiatry and Psychiatric Epidemiology*, *30*, 53–59.

Kaszniak, A. W., & Scogin, F. R. (1995). Assessment of dementia and depression in older adults. *The Clinical Psychologist*, *48*, 24–30.

Kliegl, R., Smith, J., & Baltes, P. B. (1989). Testing-the-limits and the study of adult age differences in cognitive plasticity of a mnemonic skill. *Developmental Psychology*, *25*, 247–256.

Kliegl, R., Smith, J., & Baltes, P. B. (1990). On the locus and process of magnification of age differences during mnemonic training. *Developmental Psychology*, *26*, 894–904.

Knight, B. C. (1993). Psychotherapy as applied gerontology: A contextual, cohort-based maturity-specific challenge model. In M. A. Smyer (Ed.), *Mental Health and Aging* (pp. 125–134). New York: Springer Publishing.

Korte, C. (1990). Rural elderly: Well-being, social support and social intervention. *Journal of Rural Community Psychology*, *11*, 65–82.

Korte, C., & Gupta, V. (1991). A program of friendly visitors as network builders. *The Gerontologist*, *31*, 404–407.

Kotler-Cope, S., & Camp, C. J. (1990). Memory interventions in aging populations. In E. A. Lovelace (Ed.), *Aging and cognition: Mental processes, self awareness, and interventions* (pp. 231–261). Amsterdam: Elsevier.

Kramer, A. F., Larish, J. L., & Strayer, D. L. (1995). Training for attentional control in dual-task settings: A comparison of young and old adults. *Journal of Experimental Psychology: Applied*, *1*, 50–76.

Lachman, M. E., Weaver, S. L., Bandura, M., Elliott, E., & Lewkowicz, C. J. (1992). Improving memory and control beliefs through cognitive restructuring and self-generated strategies. *Journal of Gerontology: Psychological Sciences*, *47*, 293–299.

Lang, F. R., & Carstensen, L. L. (1994). Close emotional relationships in late life: Further support for proactive aging in the social domain. *Psychology and Aging*, *9*, 315–324.

Lawton, P. M., DeVoe, R., & Parmelee, P. (1995). Relationship of events and affect in the daily life of an elderly population. *Psychology and Aging*, *10*, 469–477.

Masterpasqua, F. (1989). A competence paradigm for psychological practice. *American Psychologist*, *44*, 1366–1371.

McKitrick, L. A., Camp, C. J., & Black, F. W. (1992). Prospective memory intervention in Alzheimer's disease. *Journal of Gerontology: Psychological Sciences*, *47*, 337–343.

Moen, P. (1995). A life course approach to postretirement roles and well-being. In L. A. Bond, S. J. Cutler, & A. Grams (Eds.), *Promoting successful and productive aging* (pp. 239–256). Thousand Oaks, CA: Sage.

Munöz, R. F. Mrazek, P. J., & Haggerty, R. J. (1996). Institute of Medicine report on prevention of mental disorders. *American Psychologist*, *51*, 1116–1122.

Newman, J. P., Engel, R. J., & Jensen, J. E. (1991). Age differences in depressive symptom experiences. *Journal of Gerontology*, *46*, 224–235.

Newsom, J. T., & Schulz, R. (1996). Social support as a mediator in the relation between functional status and quality of life in older adults. *Psychology and Aging*, *11*, 34–44.

Niederehe, G. (1994). Psychosocial therapies with depressed older adults. In L. S. Scheider, C. F. Reynolds, B. D. Lebowitz, & A. J. Friedhoff (Eds.), *Diagnosis and treatment of depression in the elderly: Results of the NIH consensus development conference* (pp. 293–315). Washington, DC: American Psychiatric Press.

Okun, M. A., Olding, R. W., & Cohn, C. M. G. (1990). A meta-analysis of subjective well-being interventions among elders. *Psychological Bulletin*, *108*, 257–266.

Oxman, T. E., & Hull, J. G. (1997). Social support, depression, and activities of daily living in older heart surgery patients. *Journals of Gerontology: Psychological Sciences*, *52B*, 1–4.

Pillemer, K., & Suitor, J. J. (1996). "It takes one to help one": Effects of similar others on the well-being of caregivers. *Journals of Gerontology: Psychological Sciences*, *51B*, 250–257.

Pushkar, D., Arbuckle, T. Y., Conway, M., Chaikelson, J., & Maag, U. (in press). Everyday activity parameters and competence in elderly adults. *Psychology and Aging*.

Quayhagen, M. P., Quayhagen, M., Corbeil, R., Roth, P. A., & Rodgers, J. A.(1995). A dyadic remediation program for care recipients with dementia. *Nursing Research, 44*, 153–159.

Reich, J. W., & Zautra, A. J. (1989). A perceived control intervention for at-risk older adults. *Psychology and Aging, 4*, 415–424.

Reynolds, C. F., Frank, L., Houck, P. R., Mazumdar, S., Dew, M. A., Cornes, C., Buysse, D. J., Begley, A., & Kupfer, D. J. (1997). Which elderly patients with remitted depression remain well with continued interpersonal psychotherapy after discontinuation of anti- depressant medication? *American Journal of Psychiatry, 154*, 958–962.

Rubinstein, R. L., Lubben, J. E., & Mintzer, J. E. (1994). Social isolation and social support: An applied perspective. *Journal of Applied Gerontology, 13*, 58–72.

Salthouse, T. A. (1995). Refining the concept of psychological compensation. In R. A. Dixon & L. Bäckman (Eds.), *Compensating for Psychological Deficits and Declines* (pp. 3–21). Mahwah, NJ: Lawrence Elbaum Associates.

Salthouse, T. A. (1996). The processing-speed theory of adult age differences in cognition. *Psychological Review, 103*, 403–428.

Sandman, C. A. (1993). Memory rehabilitation in Alzheimer's disease: Preliminary findings. *Clinical Gerontologist, 13*, 19–33.

Schaie, K. W. (1983). The Seattle longitudinal study: A 21-year exploration of psychometric intelligence in adulthood. In K. W. Schaie (Ed.), *Longitudinal studies of adult psychological development* (pp. 64–135). New York: Guilford.

Schaie, K. W., & Willis, S. L. (1986). Can decline in adult intellectual functioning be reversed? *Developmental Psychology, 22*, 223–232.

Scogin, F., & McElreath, L. (1994). Efficacy of psychosocial treatments for geriatric depression: A quantitative review. *Journal of Consulting and Clinical Psychology, 62*, 69–74.

Scogin, F., & Prohaska, M. (1992). The efficacy of self-taught memory training for community-dwelling older adults. *Educational Gerontology, 18*, 751–766.

Seidenberg, M., Taylor, M. A., & Haltiner, A. (1994). Personality and self-report of cognitive functioning. *Archives of Clinical Neuropsychology, 9*, 353–361.

Smyer, M. (1995). Formal support in later life: Lessons for prevention. In L. A. Bond, S. J. Cutler, & A. Grams (Eds.), *Promoting successful and productive aging* (pp. 186–202). Thousand Oaks, CA: Sage.

Smyer, M., Brannon, D., & Cohn, M. D. (1992). Improving nursing home care through training and job redesign. *The Gerontologist, 33*, 327–333.

Staudinger, U. M., Marsiske, M., & Baltes, P. B. (1995). Resilience and reserve capacity in later adulthood: Potentials and limits of development across the lifespan. In D. Cicchetti & D. Cohen (Eds.), *Developmental psychopathology, Vol. 2: Risk, disorder and adaptation* (pp. 801–847). New York: Wiley.

Stigsdotter Neely, A., & Bäckman, L. (1993). Long-term maintenance of gains from memory training in older adults. *Journal of Gerontology: Psychological Sciences, 48*, 233–237.

Stigsdotter Neely, A., & Bäckman, L. (1995). Effects of multifactorial memory training in old age: Generalizability across tasks and individuals. *Journal of Gerontology: Psychological Sciences, 50B*, 134–140.

Sulman, J., & Wilkinson, J. (1989). An activity group for long-stay elderly patients in an acute care hospital: Program evaluation. *Canadian Journal on Aging, 8*, 34–50.

Teri, L., Curtis, J., Gallagher-Thompson, D., & Thompson, L. W. (1994). Cognitive-behaviour therapy with depressed older adults. In L. S. Schneider, C. F. Reynolds, B. D. Lebowitz, & A. J. Friedhoff (Eds), *Diagnosis and treatment of depression in the elderly: Results of the NIH consensus development conference* (pp. 279–291). Washington, DC: American Psychiatric Press.

Thompson, M. G., & Heller, K. (1993). Distinction between quality and quantity of problem-solving responses among depressed older women. *Psychology and Aging, 8*, 347–359.

Thompson, M. G., Heller, K., & Rody, C. A. (1994). Recruitment challenges in studying late-life depression: Do community samples adequately represent depressed older adults? *Psychology and Aging, 9*, 121–125.

Vaccaro, F. J. (1992). Physically aggressive elderly: A social skills training program. *Journal of Behaviour Therapy and Experimental Psychiatry, 23*, 277–288.

Verhaeghen, P., Marcoen, A., & Goosens, L. (1992). Improving memory performance in the aged through mnemonic training: A meta-analytic study. *Psychology and Aging, 7*, 242–251.

Watson, D., & Pennebaker, J. W. (1989). Health complaints, stress, and distress: Exploring the central role of negative affectivity. *Psychological Review, 96*, 234–254.

Williams, J. D., & Klug, M. G. (1996). Aging and cognition: Methodological differences in outcome. *Experimental Aging Research, 22*, 219–244.

Willis, S. L. (1989). Improvement with cognitive training: Which old dogs learn what tricks? In L. W. Poon, D. C. Rubin, & B. A. Wilson (Eds.), *Everyday cognition in adulthood and late life* (pp. 545–569). Cambridge, Great Britain: Cambridge University Press.

Willis, S. L., & Nesselroade, C. S. (1990). Long-term effects of fluid ability training in old-old age. *Developmental Psychology, 26*, 905–910.

WindRiver, W. (1993). Social isolation: Unit-based activities for impaired elders. *Journal of Gerontological Nursing, 19*, 15–21.

Yesavage, J. A. (1984). Relaxation and memory training in 39 elderly patients. *American Journal of Psychiatry, 141*, 778–781.

Yesavage, J. A., Sheikh, J. I., Friedman, L., & Tanke, E. (1990). Learning mnemonics: Roles of aging and subtle cognitive impairment. *Psychology and Aging, 5*, 133–137.

Zarit, S. H., Zarit, J. M., & Reever, K. E. (1982). Memory training for severe memory loss: Effects on senile dementia patients and their families. *The Gerontologist, 22*, 373–377.

12

POLITICAL AND SCIENTIFIC MODELS OF DEVELOPMENT*

Arnold Sameroff and W. Todd Bartko

Scientific efforts to understand child development frequently lead to different views of how these processes are organized (Sameroff, 1983). Although we believe that the results of such research could affect the lives of children, it may be that there are other paradigms that have much more salience and importance that we have tended to ignore in the past. These are the political models of development. If one needs a reminder of why these are so important, the recent political rhetoric in the United States attached to welfare reform legislation is an excellent case in point. Child advocates argued that the lives of millions of children would be jeopardized or enhanced depending on whether one supported the Democratic or Republican version of this legislation.

Scientific considerations may be relevant to understanding laboratory phenomena such as how infants find hidden objects or children learn concepts, but if one were concerned with serious child development issues, like do children grow up at all, political issues may be far more salient. There is a gap, or what may be better described as a paradigm clash between the social scientists who live in a country and the politicians who attempt to govern it. Understanding differences in paradigms or models of development is important for understanding what is happening to children today, but the relevant models are not scientific paradigms. They are political models.

A simplification for politics in the United States is to portray members of the Democratic party as seeing development being determined by social opportunities, where members of the Republican Party see it as determined by personal traits. If opportunities for good education, health, jobs and prosocial friends were available to all, then Democrats believe our social ills would disappear. In contrast, the Republican view is that those who have grit, resourcefulness, intelligence, and motivation will make it on their own and those who don't, won't. For Democrats, government has a major role at a minimum to equalize the playing field and at a maximum to tilt it in favor of those with the fewest personal resources to take advantage of what opportunities do exist. For Republicans, govern-

* Research reported here was supported by grants from the John D. and Catherine D. MacArthur Foundation Program in Successful Adolescent Development, the W.T. Grant Foundation, and the United States National Institute of Mental Health.

Improving Competence across the Lifespan, edited by Pushkar et al.
Plenum Press, New York, 1998.

ment's role should be minimized to let each individual achieve according to their innate abilities. The issue for scientists is to translate these political models into answerable empirical questions. Does research on child development address such issues? A simplified translation of political differences corresponds fairly well with the simplified scientific debate over the nature–nurture question.

In order to do an adequate assessment of the relation between characteristics of the individual and characteristics of the environment measures of both must be incorporated in developmental studies. We have been engaged in a number of such studies that may shed some light on the model issue. Although each is flawed in some way, together they offer a perspective on issues at hand in terms of where we stand now and where we need to go in the future.

DEMOCRATIC MODEL OF DEVELOPMENT

To examine the Democratic model we need to address the issue of how an individual's opportunity structure affects his or her developmental outcomes. Despite the nominal interest of developmentalists in the effects of the environment, the analysis and assessment of context has usually been oversimplified either through the sole use of proximal indicators of mother–child interaction at one extreme or distal indicators of socioeconomic status at the other. Multidisciplinary interactions with sociologists and economists have gradually enlarged the range of environmental indicators available and, increasingly, these have been introduced into developmental research.

The development of our research in this area began with such oversimplifications and may be arriving at more appropriate degrees of complexity. Our investigations of environmental influences began with my collaborators in the Rochester Longitudinal Study. At Rochester, our primary index of the environment was socioeconomic status (SES) as measured by the Hollingshead scale. We examined the effects of the environment on child development in a longitudinal investigation of a group of children from the prenatal period through adolescence living in a socially heterogeneous set of family circumstances. The original plan of the study was focused on the idea that having a parent with a psychiatric diagnosis of schizophrenia was a major risk factor for child development, primarily through mechanisms of biological intergenerational transmission. As the study progressed we had to modify this belief as we paid increasing attention to contextual factors.

During the Rochester Longitudinal Study (Sameroff, Seifer, & Zax, 1982) we assessed children and their families during the first year, and then when the children were 2½, 4, 13, and 18 years of age. At each time point we evaluated two general indicators of developmental status — the child's cognitive and social–emotional competence.

When we examined our data we found much support for an environmental hypothesis that differences in family social status would produce differences in child behavior. Children from the poorest families in our sample exhibited the poorest development starting in the second year of life when advances in communication abilities permitted them to be more influenced by differences in cognitive stimulation from their social surround (Sameroff & Seifer, 1983).

But as has been often noted, social status per se does not have a direct effect on development. It must be mediated through more proximal mechanisms if it is to affect the child. To better understand the role of SES, more differentiated views of environmental influences needed to be taken. We had to discover what was different about the experience of children raised in different economic environments.

ENVIRONMENTAL CONDITIONS AS DEVELOPMENTAL RISKS

Although SES is one of the best single variables for predicting children's cognitive competence, and an important, if less powerful predictor of social–emotional functioning, it operates at many levels of the ecology of children. It impacts on parenting, parental attitudes and beliefs, their family interactions, and many institutions in the surrounding community. From the data available in the Rochester study we searched for a set of variables that were related to economic circumstance but not the same as SES.

From the 4, 13, and 18-year assessments of the children in the Rochester study we chose a set of 10 environmental variables (Sameroff, Seifer, Barocas, Zax, & Greenspan, 1987). These definitions were: (a) a history of maternal mental illness; (b) high maternal anxiety; (c) parental perspectives that reflected rigidity in the attitudes, beliefs, and values that mothers had in regard to their child's development; (d) few positive maternal interactions with the child observed during infancy; (e) minimal maternal education; (f) head of household in unskilled occupations; (g) disadvantaged minority status; (h) single parenthood; (i) stressful life events; and (j) large family size.

When we compared the high-risk and low-risk group for each variable separately, we found, indeed, that each of these variables was a risk factor. For both the cognitive and mental health outcomes, the low risk group had better scores than the high risk group. Most of the differences were of moderate effect size, enough to demonstrate the effects for group comparisons but certainly not enough to specify which individuals with the risk factor would have an adverse outcome.

ACCUMULATING RISK FACTORS

In the Rochester Study it was the rare child with only one risk factor that ended up with a major developmental problem. But what would be the result if a comparison was made between children growing in environments with many risk factors compared to children with very few? We noted that Parmelee (Parmelee & Haber, 1973) and Rutter (1979) had argued from an epidemiological perspective that it was not any particular risk factor but the accumulated number of risk factors in a child's background that led to developmental problems. Consequently, we created a multiple risk score that was the total number of risks for each individual family.

When these multiple risk scores were related to the child's intelligence and mental health, major differences were found between those children with few risks and those with many. On an intelligence test, children with no environmental risks scored more than 30 points higher than children with seven to nine risk factors, a difference of 2 standard deviations. Even when we corrected for the contribution of mother's IQ scores, there was still more than a standard deviation difference. On average, each risk factor reduced the child's 4-year IQ score by four points.

A similar but less powerful effect was found when high and low risk children were compared on their 4-year social–emotional outcomes. Where our multiple risk variables had explained 50% of the variance in the intellectual outcome, only 25% of the variance in the social competence outcome was explained. Yet, it is clear that the effect of combining the 10 risk variables into a single measure was to strongly accentuate the differences noted when the risks were evaluated separately. As the number of risk factors increased, performance decreased for children at 4 years of age (Sameroff, Seifer, Zax, & Barocas, 1987).

CONTINUITY OF ENVIRONMENTAL RISK

Because of the potent effects of our multiple risk index at 4 years, we calculated new multiple environmental risk scores for each family based on their situation 9 and 14 years later. To our surprise there were very few families that showed major shifts in the number of risk factors across the 14-year intervening period. Between the pre-school period and adolescence, the factor that showed the most improvement was maternal education, where the number of mothers without a high school diploma or equivalent decreased from 33 to 22% of the sample. The risk factor that increased the most was single parenthood with the number of children being raised by their mothers alone increasing from 24 to 41%. In the main, however, there was little change in the environments of the children in our sample.

The effects of multiple risk were not restricted to the 4-year behavior of the Rochester children. When they were 13 years old and again when they were 18 we recalculated their multiple risk scores and examined their relation to contemporary IQ and mental health measures. Again we found strong linear relations where the more environmental risk factors, the worse the outcomes (Sameroff, Seifer, Baldwin, & Baldwin, 1993).

SECULAR TRENDS

The thrust of a contextual analysis of developmental regulation is not that individual factors in the child are non-existent or irrelevant, but that they must be studied in a context larger than the single child. The risk analyses discussed so far have implicated parent characteristics and the immediate social conditions of family support and life event stress as important moderators of healthy psychological growth in the child. To this list of risks must be added changes in the historical supports for families in a given society. The importance of this added level of complexity was emphasized when we examined secular trends in the economic well-being of families in the United States.

At 4-years we had divided the Rochester sample into low, medium, and high risk groups. We found that 22% of the high-risk group with four or more risks had IQs below 85 whereas none of the low-risk sample did. Conversely, only 4% of the high-risk sample had IQs above 115, but 59% of the low-risk group did.

After the 13-year assessment we made the same breakdown into high, medium and low-risk groups and examined the distribution of IQs within risk groups. Again we found a preponderance of low IQ scores in the high risk group and a preponderance of high IQ scores in the low risk group indicating the continuing negative effects of an unfavorable environment. But strikingly, the number of children in the high-risk group with IQs below 85 had increased from 22% to 46%, more than doubling.

If our analysis was restricted to the level of the child and family, we would hypothesize that high-risk environments operate synergistically to further worsen the intellectual standing of these children during the period from pre-school to adolescence, placing them in a downward spiral of increasing incompetence. An alternative hypothesis is that society itself was changing during the nine years between the Rochester study 4- and 13-year assessments.

In a government report (Passel, 1989) it was found that between the years 1973 to 1987, the period we were doing this study, the average household income of the poorest fifth of Americans fell 12% while the income of the richest fifth increased 24%. As the income disparity between rich and poor increased in society at large, so did the disparity between the number of children with high and low intelligence scores in our research study.

The success of the Rochester study was that we were able to gain a greater understanding of the many ways in which environments affected children. In Rochester the Democratic model of development was strongly supported. However, there was a problem of the study that limited its generalizability. The sample had a large overrepresentation of families where a parent had a psychiatric diagnosis. Would we find the same effects of multiple risks in a sample more representative of the general community? Would the Democratic model hold up in another major city?

THE PHILADELPHIA STUDY

Another opportunity to examine the effects of multiple environmental risks on child development was provided by data emerging from a study of adolescents in a large sample of Philadelphia families (Furstenberg, Cook, Eccles, Elder & Sameroff, in press). We interviewed mothers, fathers, and children in close to 500 families where there was a youth between the ages of 11 and 14. The sample varied widely in socioeconomic status and racial composition. There were 64% African-Americans, 30% non-Hispanic whites, and 6% other groups, primarily Puerto Rican families.

Other studies of multiple risk factors, like the Rochester Longitudinal Study, were important in demonstrating the power of such analyses but did not use explicitly an ecological model to identify domains of risk. Typically, there was a selection process in which the risks were chosen from the available measures already in the data set of the study. In the Philadelphia project we took a more conceptual approach in designing the project so that we had environmental measures at a series of ecological levels.

For our analyses of environmental risk we examined variables within systems that affected the adolescent, from those microsystems (Bronfenbrenner, 1979) in which the child was an active participant to those systems more distal to the child where any effect had to be mediated by more proximal variables. We made a distinction between the characteristics of systems that were theoretically independent of the child and those in which the child was an active participant. For example, the family system was subdivided into management processes where it is difficult to determine if the behavior is influenced more by the parent or the child, such as behavioral control, and structural variables, such as marital status and household density, that were relatively independent of the child (see Table 1).

The risk factors were placed in six groups reflecting different ecological relations to the adolescent (Sameroff et al., in preparation). We selected 20 variables to serve as risk factors, twice as many as in the Rochester study. Our intention was to be able to have multiple factors in each of our six ecological levels. *Family Process* was the first grouping and included variables in the family microsystem that were directly experienced by the child. These included support for autonomy, behavior control, parental involvement, and family climate. The second grouping was *Parent Characteristics* which included the mother's mental health, sense of efficacy, resourcefulness, and level of education. This group included variables that influenced the child but, generally speaking were less influenced by the child. The third grouping was *Family Structure* that included the parents' marital status, and socioeconomic indicators of household crowding and receiving welfare payments. The fourth grouping was *Family Management of the Community* and was comprised of variables that characterized the family's management of its relation to the larger community as reflected in variables of institutional involvement, informal networks, social resources, and adjustments to economic pressure. The fifth grouping, *Peers*, included indicators of another microsystem of the child, the extent to which the youth was associ-

Table 1. Risk variables in Philadelphia study

Domain	Variable
Family process	
	Support for autonomy
	Discipline effectiveness
	Parental investment
	Family climate
Parent characteristics	
	Education
	Efficacy
	Resourcefulness
	Mental health
Family structure	
	Marital status
	Household crowding
	Welfare receipt
Management of community	
	Institutional involvement
	Informal networks
	Social resources
	Economic adjustment
Peers	
	Prosocial
	Antisocial
Community	
	Neighborhood SES
	Neighborhood problems
	School climate

ated with prosocial and antisocial peers. *Community* was the sixth grouping representing the ecological level most distal to the youth and the family. It included a census tract variable reflecting the average income and educational level of the neighborhood the family lived in, a parental report of the number of problems in the neighborhood, and the climate of the adolescent's school.

In the Philadelphia study, in addition to the larger number of ecological variables, we had a wider array of assessments available for interpreting developmental competence. The five outcomes that we thought would characterize successful adolescence were parent reports of adolescent *Psychological Adjustment* on a number of mental health scales, youth reports of *Self-Competence* measures and *Problem Behavior* with drugs, delinquency, and sexual behavior, and combined youth and parent reports of *Activity Involvement* in sports, religious, extracurricular, and community projects and *Academic Performance* as reflected in grade reports reported by the parent and the adolescent.

Identifying Risks

As in the Rochester Longitudinal Study, once we had determined a representative list of potential risk factors at different levels of the child's ecology, we had to assess whether each of these variables was indeed a risk factor. We used two criteria for identifying each risk factor. The first was that the variable was correlated with one of our five outcome variables and the second was that adolescents in families that had the risk factor did significantly worse on at least one of the outcomes than adolescents in families without that environmental risk. For those variables that met the criteria, we chose a cutoff score to optimize the difference between the outcomes for adolescents with the risk factor and those without. In general, the cutoff separated about 25% of the sample as a high-risk group from the remaining 75% defined as low risk.

The first important conclusion was that, indeed, we were successful in identifying risk factors. We found that there were risks at every ecological level associated with child outcomes. It was not only the parent or the family that had an influence on child competence, but also the peer group, neighborhood and community together with their interactions with the family. Some of the variables were risks for each of our five outcomes. These included lack of support for autonomy, a negative family climate, and few prosocial peers. At the other extreme were variables that affected only a few outcomes such as having parents who lacked education and resourcefulness, single marital status and much economic adjustment, a lack of informal networks, and low census tract socioeconomic status.

Many risk factors have been identified in previous research that used only a single adolescent outcome such as delinquency (Stouthamer-Loeber, Loeber, Farrington, Zhang, van Kammen, & Maguin, 1993). To examine the generalizability of risk factors requires that there be multiple outcomes in the study. In the Philadelphia study we found that the pattern of relations between ecological variables chosen as our risk factors and adolescent behavior was different for each outcome. On the one hand, academic performance, psychological adjustment, and problem behavior were related to risks at every ecological level. On the other hand, the correlates of self-competence and activity involvement presented two more limited and contrasting pictures. Activity involvement was strongly related to family management of the community and community characteristics, whereas self-competence was unrelated to either. In contrast, family structure played a significant role in adolescent self-competence but not in their activity involvement.

As in the Rochester study, when the differences between high and low risk groups were examined for each individual risk factor, the effect sizes were small or moderate, rarely exceeding ⅓ of a standard deviation. But as in Rochester we could ask the question of what would be the consequence on adolescent competence if the youth experienced a number of these risk factors? Moreover, what would be the increase in predictive efficiency if we used a cumulative risk score as our predictor for adolescent success?

Multiple-Risk Scores

Multiple environmental risk scores were calculated for each adolescent. The resulting range was from a minimum of 0 to a maximum of 13 with a mode of 5 out of a possible 20 risk factors. Only one family had no risk factors. For the multiple risk analysis we wanted to have an adequate sample size in each group so we combined the single family with no risk factors with the one-risk factor group and at the other extreme the 65 families with 9 or more risk factors into a single group. When the five normalized adolescent outcome scores were plotted against the number of risk factors a very large decline in outcome was found with increasing risk.

As can be seen in Figure 1 the maximum effect of cumulative risk was on psychological adjustment and academic performance, with a difference of more than one and a-half standard deviations between adolescents with only one risk factor compared to those with nine or more. The smallest effects were for the youths' report of self competence and activity involvement where the difference was less than a standard deviation.

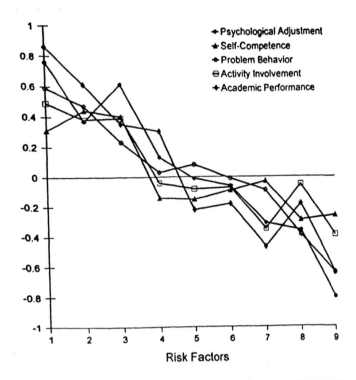

Figure 1. Relation of five youth outcomes to multiple risk scores in Philadelphia study.

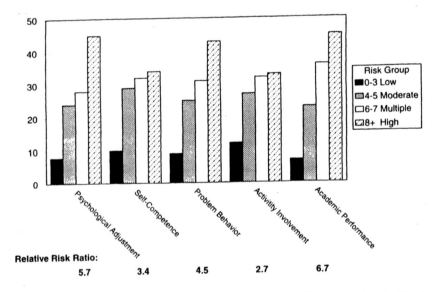

Figure 2. Percent of adolescents with poor outcomes (in the lowest quartile of five outcomes) in the Philadelphia study separated into four different risk groups. Odds are calculated as the ratio between percent of youth in the lowest quartile in the high risk and low risk groups.

Odds-Ratio Analysis

Whether our cumulative risk score meaningfully increases our predictive efficiency can be demonstrated by an odds-ratio analysis. We could compare the odds of having a bad outcome in a high risk versus a low risk environment. For the typical analysis of relative and attributable risk the outcome variable is usually discrete, succumbing to a disease or disorder. For our sample of young adolescents, there were few discrete negative outcomes. They were generally too young to have many pregnancies or arrests and the rate of academic failure was not particularly high. We had to artificially create bad outcomes by making cut scores in our outcome measures. We dichotomized each of the five outcomes by making a cut at the 25th percentile for worse performance. These were the 25% of adolescents who were doing the most poorly in7 terms of mental health, self-competence, problem behavior, activity involvement, or academic performance. To simplify the report we examined the relation between these bad outcomes and adolescent environmental risk scores subdivided into four multiple risk groups: a low risk group defined as 3 or less, two moderate risk groups of 4–5 and 6–7 risks, and a high risk group with 8 or more (see Figure 2).

The relative risk in the high risk group for each of the bad outcomes was substantially higher than in the low risk group. The strongest effects were for academic performance where the relative risk for a bad outcome increased from 7% in the low risk group to 45% in the high risk group, an odds ratio of 6.7 to 1. The weakest effect was for activity involvement where the relative risk only increased from 12 to 33%, an odds ratio of 2.7 to 1. In some sense this is not unexpected because where everyone would agree that academic failure and poor mental health are bad outcomes, there might be some dispute whether an adolescent's desire not to participate in the scouts, religious activities, or sports reflects a lack of competence. In any case, for the important cognitive and social–emotional outcomes of youth there seem to be powerful negative effects of the accumulation of environmental risk factors.

Republican Model of Development

At this point it would seem that environmental factors play a major role in the achievement of children in our society, both in academic accomplishment and in the mental health that characterizes good social and emotional functioning. But it's time to consider the Republican model. Can differences in achievement be better explained by individual characteristics of the child. Is it possible that despite social adversity those children with high levels of personal resources, what is coming to be called psychological capital (Coleman, 1988), are able to overcome minimal resources at home and in the community to reach levels of achievement comparable to children from more highly advantaged social strata.

The psychological constructs that most approximate Republican "true grit" are resourcefulness and efficacy. In the Philadelphia study we were able to measure this construct with a set of questions asked of the parent and child about the youth's capacity to solve problems, overcome difficulties, and bounce back from setbacks. We divided the sample into high and low efficacy groups and looked at their adolescent outcomes. Indeed, high efficacious youth were more competent than those with low efficacy on our measures of adolescent competence. A sense of personal resourcefulness did seem to pay off.

But what happens to this effect when we take environmental adversity into account? When we matched high and low efficacy children for the number of environmental risk factors, the difference in general competence between youth in the high and low environmental risk conditions was far greater than that between high resourceful and low resourceful groups (see Figure 3). High efficacious adolescents in high risk conditions did

Figure 3. Efficacy by competence. Relation of multiple risk to outcome for Philadelphia youth in high and low resourcefulness groups.

worse than low efficacious youth in low risk conditions. It may not be a surprise to learn that the ineffective offspring of advantaged families may have a much easier ride than resourceful multi-risk children (Sameroff, Bartko, & Eccles, in preparation).

We did the same analysis using academic achievement as an indicator of competence and examined whether their good work at school was related to better mental health, more engagement in positive community activities, and less involvement in delinquent problem behavior. Again for every outcome, high academic achieving adolescents in high risk conditions did worse than youth with low school grades in low risk conditions.

One of the major weaknesses in the Philadelphia study is that it is cross-sectional. Causal modeling is impossible unless one has longitudinal developmental data, and difficult even then. The Rochester study did have a series of developmental assessments which permitted a longitudinal view of the contribution of individual factors to developmental success.

From the Rochester data collected during the first year of life we created a multiple competence score for each child during infancy that included 12 factors. These were neonatal behavioral test scores, easy temperament scores, and developmental quotients. We then divided the sample into low, medium and high competence groups of infants and examined as outcomes their 4-year IQ and social-emotional functioning scores.

We found no relation between infant scores and 4-year IQ, especially when compared to the effects of contemporaneous infant environmental multi-risk scores. Similarly, there was no relation between infant multiple risk scores and the global rating of 4-year social competence.

However, there is a general feeling that infant developmental scales may be weak predictors because they assess different developmental functions than are captured by later

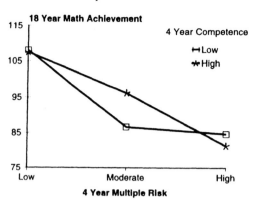

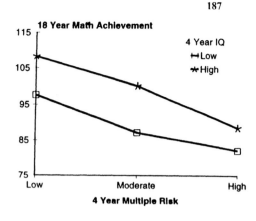

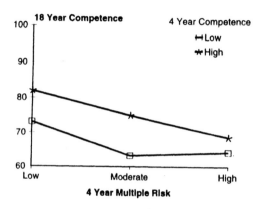

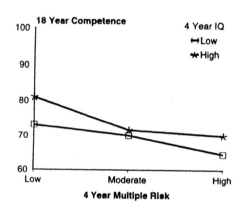

Figure 4. Relation of 4-year multiple risk to 18-year math achievement or social emotional competence for groups of children who were high and low on 4-year social–emotional competence or IQ.

cognitive and personality assessments. Perhaps, if we move up the age scale we may find that characteristics of these children at 4 years of age contribute to adolescent achievements at our 18-year assessment.

We did not have a specific measure of resourcefulness at 4 years but we did have our global measures of child social competence and IQ. We divided the 4-year-olds into high and low social competence groups and high and low IQ groups. We then compared these groups on how they did at 18 years on their mental health and measures of school achievement (see Figure 4). More resourceful children did better on average than less resourceful children but as in the Philadelphia data, when we controlled for environmental risk, the differences between children with high and low levels of human capital paled when compared to the differences in performance between children in high and low-risk social environments. In each case, high competent children in high risk environments did worse than low competent children in low risk environments.

Perhaps 4-year competence is still too ephemeral to resist the negative consequences of adverse social circumstance. Would competent children at 13 years succeed where competent children at 4 years had failed? How would they stack up at our 18 year assessments of mental health and school achievement? At 13 years we did have an index of resourcefulness in scores on an internal sense of locus of control. We divided the sample

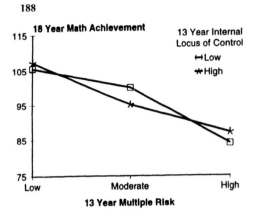

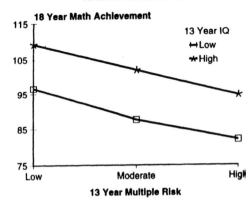

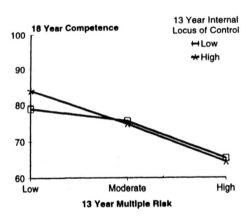

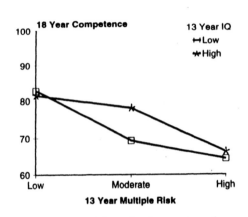

Figure 5. Relation of 13-year multiple risk to 18-year math achievement or social emotional competence for groups of children who were high and low on 13-year internal locus of control or IQ.

of children into a high and low internal locus of control group and a high and low intelligence group and examined their 18-year behavior.

Again, in each case when we controlled for environmental risk we found that the highly competent children in personality or intelligence did far less well than we would expect. Those groups of children with high levels of human capital living in conditions of high environmental risk did worse than similar groups in low risk conditions, but even more to the point, did worse than low competent children in low risk environments (see Figure 5).

The Republican model of development may hold on a level playing ground where children are faced with similar levels of environmental advantage or adversity, but this is not yet true in the United States, where environmental inequality is generally pervasive. The negative effects of a disadvantaged environment seem to be more powerful contributors to child achievement at every age than the personality characteristics of the child.

DEMOCRATS VS. REPUBLICANS

The results from the two aforementioned longitudinal studies, where some rough comparisons were made between the effects of environmental circumstance and individual competence on developmental achievements, would seem to support the Democratic model of development over the Republican one. Environments play a critical role in supporting child development and to the extent that we wish to improve the life accomplishments of our children we need to strongly support efforts to improve those environments.

But at the beginning of this chapter a question was raised about the difference between scientific and political models of development. Examining this difference could be considered merely a semantic exercise, if it did not have major implications for finding a workable solution for many of our current social problems.

What would be defined as a difference in scientific models would be a difference in level of complexity, as for example the differences between mechanistic and organismic paradigms (Sameroff, 1983). Where in the mechanistic model, all more complicated systems can be reduced to a single set of underlying components, in the organismic, each part must be considered in relation to an organizing holistic structure. In contrast, differences in political models are usually between two views at the same level of paradigm complexity. For the most part, politicians from both American parties use the same level of complexity, and unfortunately, it is a simplistic one. Life's problems have single solutions. The Republican view is that if we simply unfetter our citizens from government constraint they will be free to achieve wealth, health, and happiness. The Democratic path to the same goal is that all we have to do is to provide more resources to families and they will achieve happiness. The paradigm that both sets of politicians try to avoid is that life is more complicated.

We have presented data to show that an individually oriented view of development is generally overpowered by environmental considerations. However, we needed to add a further amplification of the contextualist position. Environments are not simple. If one were looking for an easy way to improve them, one would be at a great loss in the face of the complexity of ecological models.

Some natural scientists, and even some politicians, make fun of social scientists for presenting life as more complicated than they believe is necessary. This is based on the mechanistic notion that life is indeed simple. Perhaps if we examine some simple biological process we will be able to see how easy it is to describe. The regulation of blood pressure would seem to be such a simple function, but unfortunately, 200 subsystems are engaged in this process. The role of each component has been empirically documented. In comparison the small sets of 10 or 20 risk factors we use to examine environmental effects are quite simple. But there are many who wish to make it even simpler.

In the Philadelphia study we did a further set of analyses to illuminate some problems in the Democratic position. We examined the effect of some single risk factors that social activists have been very concerned about, income level and marital status. Although one would think that these factors should have powerful effects on the fate of children, the differences in competence disappeared when we controlled for the number of environmental risk factors in each family. First we split our sample of families into those with high, middle, and low income levels. Then we split the sample into groups of children living in two-parent vs. single-parent families. The outcome lines seem to be completely overlapping when we compare groups of children with the same number of risk factors raised in richer or poorer families or families with one or two parents (see Figures 6 and 7).

What our analyses of the data reveal is that it is not single environmental factors that make a difference but the constellation of risks in each family's life. The reason that income and marital status seem to make major differences in child development is not because they are overarching variables in themselves, but because they are strongly associated with a combination of other risk factors. Table 2 depicts the percentages of families with different income levels and families with one- or two-parent homes in groups with different numbers of risk factors. One can see in the last row of Table 2 that where 44% of poor children lived

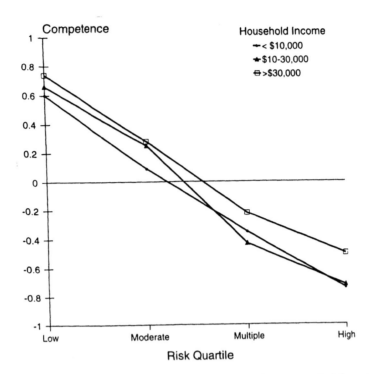

Figure 6. Income and multiple risk. Relation of risk scores to a composite measure of adolescent competence in groups with different amounts of total household income.

in high-risk families, only 6% of affluent children did. Similarly, where 22% of single parent families lived in high risk social conditions, only 15% of two-parent families did.

What these analyses reveal is that income or marital status taken alone may have statistically significant effects on adolescent behavior, but that these differences are small in comparison with the effects of the accumulation of multiple negative influences that characterize our high risk groups. The overlap in outcomes for youth in high and low income families, and in single and two-parent families is substantial for any and all psychological outcomes. There are many successful adults who were raised in poverty and unsuccessful ones who were raised in affluence. There are many healthy and happy adults who come from broken homes, and there are many unhappy ones who were raised by two parents.

Table 2. Percent risk by income and marital status*

	Income level			Family structure	
	$10,000	$10–30,000	>$30,000	Single parent	Two parent
N	113	208	142	250	235
Risk group					
Low	13%	35%	60%	34%	54%
Moderate	17%	22%	23%	25%	22%
Multiple	26%	28%	11%	19%	17%
High	44%	15%	6%	22%	7%

*Percent of Philadelphia study families in different risk groups with household incomes less than $10,000 and $30,000, and greater than $30,000 and in single-parent vs. two-parent homes.

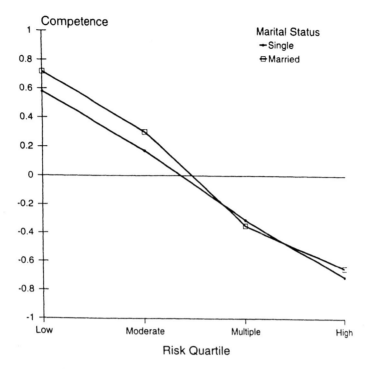

Figure 7. Marital status and multiple risk. Relation of risk scores to a composite measure of adolescent competence in single- vs. two-parent families.

The important implication is that a focus on single characteristics of individuals, like resourcefulness or intelligence, or families, like welfare or marital status, can never explain more than a small proportion of variance in normal behavioral development involving a wide variety of environments. But major differences do emerge when comparisons are made between groups of children reared in conditions with many risk factors and those with only a few. To truly appreciate the determinants of competency requires attention being paid to a broad constellation of ecological factors in which these individuals and families are embedded.

CONCLUSIONS

What we have described are the results of a research agenda for understanding why all children do not grow up to be healthy, happy, productive citizens. Republicans would argue that children do poorly in conditions of adversity because they do not have individual characteristics that would promote resilience, overcome challenge, and eventuate in productive work and family life. In contrast is the Democratic position that in conditions of poverty, opportunities do not exist for positive development even if the child does have excellent coping skills.

What is clear from our research is that there is no emergent simplification on either the environmental or constitutional side that can explain how successful development occurs or how development can be changed. Single factors can be potent in destroying systems — an earthquake can destroy a city, or a gunshot can destroy a child. But single factors cannot create a child or any other living system. At the biological level 100,000

genes are required to transform an egg cell into an adult human body, each gene expressing itself in precise degrees at precise times in precise locations. It may take far more than
100,000 events to produce the complex psychological functioning of the adult human.

What we have demonstrated in our work is that no one of these events alone deter-
mines outcomes. Each makes its small contribution to the developing child, positively for
the fortunate, negatively for those less lucky. In the data analyses we described, even
using sets of 20 risk factors, heavily biased on the environmental side, we rarely explain
more than a quarter of outcome variance in large populations. But by using such sets of
multiple variables we may do a better job of identifying those youth most in danger
because of the large number of risk factors they are experiencing, or helping to create.
Perhaps we can maximize the efficiency of intervention efforts when we realize that it is
not being poor alone, or living in a bad neighborhood alone, or having a single parent
alone that places children at risk, but rather the combination of these factors that saps the
lives of families.

While political models of development may dominate the rhetoric devoted to
explaining why children succeed or fail, it is equally clear that neither will offer a strategy
that will substantially change the lives of children living in contemporary society. Within
science, metaphors of complexity are coming to dominate thinking in most current disciplines, with the accompanying need to attend to multiple levels of analysis, multiple value
systems, and multiple life styles. Hopefully, within such a framework more adequate solutions to the problems of child development may emerge.

REFERENCES

Bronfenbrenner, U. (1979). *The Ecology of Human Development*. Cambridge: Harvard University Press.

Coleman, J. (1988). Social capital in the creation of human capital. *American Journal of Sociology, 94*, S95–S120.

Furstenberg, F. F., Jr., Cook, T. D., Eccles, J., Elder, G. H., Jr., & Sameroff, A. (in press). *Managing to make it:
     Urban families and adolescent success*. Chicago: University of Chicago Press.

Parmelee, A. H., & Haber, A. (1973). Who is the "risk infant"? In H. J. Osofsky (Ed.), *Clinical Obstetrics and
     Gynecology, 16*, 376.

Passell, P. "Forces In Society And Reaganism, Helped By Deep Hole For Poor." *New York Times*, 16 July 1989, p. 1,
     20.

Rutter, M. (1979). Protective factors in children's responses to stress and disadvantage. In M. W. Kent & J. E. Rolf
     (Eds..), *Primary prevention of psychopathology (Vol. 3): Social competence in children* (pp. 49–74). Hanover,
     NH: University Press of New England.

Sameroff, A. J. (1983). Developmental systems: Contexts and evolution. In W. Kessen (Ed.). *History, theories, and
     methods*. Volume I of Ph. H. Mussen (Ed.), *Handbook of Child Psychology* (pp. 237–294). New York:
     Wiley.

Sameroff, A. J., Bartko, W. T., & Eccles, J. (In preparation). The relation between individual and family factors as
     determinants of successful adolescent development.

Sameroff, A. J., & Seifer, R. (1983). *Sources of continuity in parent–child relationships*. Paper presented at the
     meeting of the Society for Research in Child Development, Detroit.

Sameroff, A. J., Seifer, R., Baldwin, A., & Baldwin, C. (1993). Stability of intelligence from preschool to adoles-
     cence: The influence of social and family risk factors. *Child Development, 64*, 80–97.

Sameroff, A. J., Seifer, R., Barocas, R., Zax, M., & Greenspan, S. (1987). Intelligence quotient scores of 4-year-old
     children: Social environmental risk factors. *Pediatrics, 79*, 343–350.

Sameroff, A. J., Seifer, R., & Zax, M. (1982). Early development of children at risk for emotional disorder. *Mono-
     graphs of the Society for Research in Child Development*, 47:(Serial No. 199).

Sameroff, A. J., Seifer, R., Zax, M., & Barocas, R. (1987). Early indicators of developmental risk: The Rochester
     Longitudinal Study. *Schizophrenia Bulletin, 13*, 383–393.

Stouthamer-Loeber, M., Loeber, R., Farrington, D. P., Zhang, Q., van Kammen, W., & Maguin, E. (1993). The
     double edge of protective and risk factors for delinquency: Interrelations and developmental patterns.
     Development and Psychopathology, 5, 683–701.

AFTERWORD

Improving Competence — Is It Ethical to Consider Cost?

Frederick Lowy and Anna-Beth Doyle

Improving competence across the lifespan has been an objective of human kind since recorded history. People have tried to function more effectively as they sought to secure the necessities of life, create and adapt to social structures, and achieve personal growth, both in terms of coping with dysfunction in the context of harsh physical and psychological environments, and in terms of striving to maximize human potential in favorable circumstances. One might say that the entire process of creating civilizations is the product of humanity's collective success in improving competence. In general, enhancing competence has also been regarded as a "good" for individuals and is rewarded in many ways. Those who assist others to improve their competence are also rewarded including those who provide professional services, those who teach and those who create knowledge that will lead to greater human competence. Indeed, many regard the creation and application of knowledge that will improve human competence to be an ethical imperative.

Throughout history, individuals and societies have applied resources towards improving competence and have regarded such application as an investment that will, in future, yield desirable outcomes. There has always been tension between the allocation of resources for improving competence in the long run and the consumption of resources to satisfy immediate needs and wants. The balance between these competing expenditures has varied in different societies and different periods of time and, of course, within societies, both individuals and families differ in their choices.

NOW WHAT HAS ETHICS TO DO WITH ALL THIS, AND WHERE DOES COST COME IN?

Ethics can be defined broadly as that field of inquiry concerned with value judgements concerning the rightness or goodness of decisions. Since Aristotle, the challenge has been to delineate right and good actions from those that are not right and good, nowadays often called "unethical." Ethics of resource utilization has become a major topic in

Improving Competence across the Lifespan, edited by Pushkar *et al.*
Plenum Press, New York, 1998.

20th century moral philosophy. It is not a subject that had previously been prominent in the history of ethical discourse.

Two developments are responsible for this new addition. The first is the enormous expansion of human potential. The second is the consolidation in our century of the late 18th century notion of collective entitlement that found expression in the American and French revolutions. These important developments deserve brief elaboration.

It is hardly necessary to note once again that more progress has occurred in basic and applied knowledge in the past half century than in all the eons of sentient human existence. Our understanding of the natural world and of the determinants of the human condition has expanded vastly. As a result we have considerable information about how the physical and psychological environment can be modified to protect human development and a growing appreciation of the elements required to advance individual biopsychosocial competence. Some advances of these kinds are the subjects of this book. In short, we have unprecedented capacity to alter for the betterment of the human condition.

Parallel with this expansion of potential has been the sociopolitical acceptance that certain human needs can be regarded as rights, the fulfilment of which is a collective societal obligation. Among those rights/obligations are not only a safe physical environment — such things as clean water, safe food, protection from known toxins, etc. — but also the psycho-social-educational milieu that favours optimal human intellectual and affective development. This has led to the publicly-supported institutions that we now take for granted — child protection laws and agencies, good schools, public libraries, recreational facilities, and so on. Where special needs have been identified, additional societally funded facilities and programs have been made available, such as day-care centres, head start programs, special educational facilities and the like. Implicit in all this are the assumptions that knowledge that becomes available about the determinants of human potential should, ideally, be translated into applied programs to enhance competence and these should be made available to all persons they would benefit.

It is precisely this ideal of universal entitlement to a rapidly expanding number of probably beneficial interventions that raises the ethical issues around resource allocation. Needless to say, if resources were infinite, everything any person might need (and, perhaps, anything any person might want) could and, arguably, should be made available. But we live in a world of finite resources within which a host of opportunities and expectations compete for recognition. The ethical challenge is to identify the line that must be drawn, at any point in time, between those expenditures that we regard as necessary in a caring, civilized society and those that, reluctantly, we define as discretionary and therefore can be available to those who can afford them but not to everyone else.

How do we approach this challenge? One might think that easy answers would be at hand at least in the area of mental health care. In most developed countries, universal access to needed health care is either a reality or an ideal about which there is broad consensus. It is accepted that protection from mental illness, treatment of disease and reduction of disability should be available to all because their absence threatens the maintenance and improvement of human competence. Yet even in countries like Canada where this ideal has been translated into partial reality (never mind that the affordability of optimum health care for everyone is now in serious question) it is becoming increasingly clear that difficult decisions are necessary that are social–philosophical and moral rather than strictly scientific and medical.

Take, for example, the problem of abnormally short stature. It is clear that people who are extremely short because of a deficiency of growth hormone due to an unfortunate genetic endowment are unlikely to reach psychosocial potential. Until the isolation of growth hormone and, more recently, its wider availability through genetic engineering techniques, dwarfs struggled to find a self-respecting place in society. We doubt that anyone would deny growth hormone treatment to, say, a pituitary dwarf child who without such treatment will attain a height of no more than 3½ feet even though the cost of such treatment is in the thousands of dollars per year. There is probably broad agreement that national health insurance, as in Canada, or private health insurance, as in the United States, should make possible such treatment for such a child because of the undeniable enhancement of competence that will result from attaining normal stature. But what is normal stature? In the ethnically mixed societies that characterize most countries, the normal range of height, weight, musculature, etc., is broad. Yet there are social psychological studies that demonstrate, at least in our society, that tall persons on average acquire some social advantages over short persons — they get ahead faster, they do better in many sports, they have enhanced self-esteem and assertiveness. What if parents of a boy destined to be, say 5 feet 7 inches, ask his physician to arrange for growth hormone therapy so that he can attain a height of 6 feet 2 inches? What if parents in a disadvantaged neighbourhood want their son to be 7 feet 2 inches so that, as a good basketball player, he can escape his limiting environment and enhance his psycho-socio-economic development? Is it justified for society to expend limited resources to achieve greater competence for such individuals?

The growth hormone example can be viewed as a paradigm for treatments that are not, strictly speaking, illness reducing but more accurately, must be seen as competence extending. Many costly psychological interventions fall into this category. In the realm of clinical psychology and psychiatry, to be sure, it is often difficult to draw a line between frank psychopathology and personal/social ineptness (even allowing for the generous inclusivity of the Diagnostic and Statistical Manual of Mental Disorders (4th ed; DSM-IV). Nevertheless, honest psychotherapists will admit that once symptomatic improvement has been achieved, ongoing treatment usually has the aim of enhancing competence — in interpersonal relationships, expression of autonomy and assertiveness, self-actualization, etc. Should cost be considered here? And cost to whom? If the patient/client (or his/her parents) are funding the intervention it can be argued that this is simply like other competence improving societal "goods" that are being purchased privately, like skill enhancing ballet lessons, music lessons, tennis lessons or even horizon-broadening education, foreign travel, and so on. But what if society (via tax-supported medicare or premium dependant insurance) is being asked to pay for them? To what point should collective resources be expended to improve the competence of individuals? Few would quarrel, for example, with providing soft baby carriers to low income mothers, given that the carriers were found to increase the probability of their infants being securely attached, through promoting the mothers' carrying, physical contact, and maternal responsiveness (Anisfeld, Casper, Nozyce, & Cunningham, 1990). Secure child–mother emotional attachment is widely regarded as a protective factor against later problems of adjustment and this intervention is low cost.

Another, more challenging dilemma is when a highly effective intervention in one context is less effective in another context. The Perry Preschool Program, a one- or two-year preventive intervention for children living in poverty and at risk for school failure, was found to result at age 19 in participants having significantly less public assistance, delinquency, and, for girls, pregnancy, and having greater academic achievement, than matched non-participants (Schweinart & Weikart, 1988). For middle-class children, however, the five-year Brookline Early Education Project resulted in only a modest reduction in innappropriate classroom behaviour and reading difficulty (Pierson, Walker, & Tivnan, 1984).

Cost-benefit or cost-effectiveness analyses (Mrazek & Haggerty, 1994) help clarify the contributions of these interventions. For example, the Perry Preschool Project returned $3–6 for every dollar invested in terms of savings in special education, crime and welfare assistance costs (Schweinhart & Weikart, 1988), the Brookline Early Education Project less tangible results. But we are left with the question of whether it is ethical to continue the former program but not the latter.

There are many historical examples of societies that have invested resources to improve certain competences not primarily for the benefit of individuals but, rather, to achieve specific societal objectives. Examples are the enhancement of physical development and discipline of men in ancient Sparta to improve Sparta's military potential, the investment of huge and disproportional sums of public money in the former East Germany or Cuba in the development of children and teenagers with athletic potential to achieve international athletic success so as to add prestige for the country, and so on. Cost seems not to have been the major consideration in these examples.

To sharpen the question it is helpful to consider other examples that illustrate the ethical issue. These come from the health field and its boundaries, partly because it is here that cost considerations and issues of distributive justice have become especially important.

In a paper on determining "medical necessity" in mental health practice, James Sabin and Norman Daniels (1994) ask, and we quote:

> "Which kinds of mental suffering create a legitimate claim for assistance from others through health insurance? When should individuals be responsible for correcting their own deficits of happiness or well-being or for the disadvantages they suffer? And even if society concludes that an individual is entitled to assistance from others, when does this obligation fall to friends, families, or other social agencies, rather than to the health insurance system?"

A moment's reflection will establish that "medical necessity" is defined differently by different people and the differences reflect moral and political considerations as much as clinical issues. In the previous example, the parents who want prescriptions for growth hormone so that their child of normal stature can become taller, the physician who evaluates the request and the third party payer may well differ as to whether the intervention is necessary or even justified, let alone who should pay for it.

What about the musician who wants a series of prescriptions for propranolol? This drug, a beta-adrenergic blocker, acts on the sympathetic nervous system mediated somatic concomitants of anxiety. In small, safe doses, it counteracts the rapid pulse, shallow breathing, dry mouth, perspiration, gastrointestinal mobility, etc., that accompany anxiety in the face of stress. In this case we are talking about performance anxiety of the type commonly called "stage fright." Propranolol was developed as a treatment for angina, cardiac arrhythmias and hypertension. It is now at least as often used for enhancing the performance of musicians, public speakers, test takers and the like. Is the solo pianist who takes 10 mg of propranolol before each performance being treated for a social phobia, which is a societally sanctioned activity because social phobia is defined as a psychopathological disorder with a DSM-IV and an International Classification of Diseases (ICD) number? Or is the pianist unfairly using (or abusing) a drug to "artificially" enhance what might be a fine performance even without the drug? Is the pianist cheating as we consider the athlete to be whose performance is enhanced by stanazolol or other anabolic steroids? (In the 1988 Seoul Olympics, Canadian Ben Johnson, then considered the world's fastest human, lost his gold medal and his reputation because he was caught enhancing his performance through drugs.)

Clearly, judgements of the rightness or wrongness of the use of the beta-blocker or the steroid to enhance performance have to do with moral beliefs. Jacquelyn Slomka (1992), in her paper "Playing with Propranolol" points out that even drugs to relieve pain are acceptable or unacceptable because of beliefs and values that shift. In Victorian England, the use of drugs to relieve childbirth pain was seen to be contrary to the Biblical statement "in sorrow she will bring forth children" and therefore discouraged on moral grounds until Queen Victoria herself insisted on having an anesthetic for her deliveries.

Is it morally better to suffer the pain of childbirth, or struggle through a stressful public performance by "toughing it out" on your own, or to suffer through a bout of depression by sheer willpower, than to receive clinical help in these situations? The answer depends on values that are determined by time and place. It is only recently that physicians are prescribing adequate doses of narcotic drugs to persons suffering from cancer or AIDS without feeling guilty that they are doing something that is morally suspect. An ever-increasing pharmacopoeia of pleasure producing or discomfort reducing or performance enhancing drugs is becoming available as understanding of brain mechanisms and neuro-pharmacology advances.

It is enough here just to mention the selective serotonin re-uptake inhibitors (SSRIs) now being used widely to reduce shyness, increase mental activity, and to add zest when one is hit with the "blahs" — see Peter Kramer's best seller "Listening to Prozac" (1993) as an example of authors who justify the use of SSRIs for such purposes. Although they did not know Prozac or Zoloft or Paxil, Timothy Leary and Carlos Castañeda, 30 years ago in the 1960s and, for that matter, Aldous Huxley thirty years earlier in the 1930s expressed the same theme: Human experience can be enhanced by drugs so why not indulge. We also could speak of Ritalin for distractedness, the ergoloid myelates to enhance memory, Dilantin to raise IQ, oxytocins to increase female libido, testosterone to preserve youthfulness and a host of designer drugs to come.

Let us shift from drugs to psychological interventions. The Canada Health Act, which governs eligibility for free physician and hospital services in this country, requires, among other things, that services provided be "comprehensive" and "universal." Since this could be interpreted to mean entitlement for everyone to everything that might improve their situation it became necessary to define who is eligible to receive what. Each of the Canadian provinces which administer and increasingly fund this national health insurance plan, defines eligibility differently. Predictably, as funding available fails to meet demand, eligibility is being narrowed. More strenuous attempts are being made to distinguish treatment for disability from enhancement. It is relatively easy to say that plastic surgery to restore function and appearance after an automobile accident should be freely provided under the Act but cosmetic plastic surgery should not. It is much harder to determine whether dynamic psychotherapy or cognitive–behaviour therapy should be publicly funded for the purpose of increasing assertiveness for shy people, of helping an irascible person control his/her temper, or to assist the acquisition of better social skills for a lonely person. Similarly, in the field of education, we are faced with the question of whether pre-school education, of demonstrable benefit for those children at risk because of poverty and low-educational achievement of their parents, but also a major service to middle-class families with two working parents, should be universally available and publically funded. The Quebec government is currently on the horns of this dilemma, having announced a universal program but having delayed its implementation to date because of lack of funds.

In most Canadian provinces, health care expenditures account for one-third of the budget thereby squeezing expenditures for other desirable government programs (such as support for universities). In the United States the fear of opening a bottomless pit is one of the factors that impede more favourable consideration of publicly funded universal health care.

When, then, is it ethically justified for cost to be a major consideration in deciding medical and educational intervention and when should this not be the case? (we do not believe there is any circumstance in which cost considerations can be totally and absolutely disregarded.) Since many of the symptoms of mental illness lie along a continuum of everyday normal experiences and it has been demonstrated that the availability of mental health insurance coverage increases demand for services it is important not only to delineate what interventions should be made available for what condition but, as Sabin and Daniels (1994) point out, what is the ultimate goal of the intervention. They construct three models that, they believe, underlie decisions actually made by mental health providers in settings where "medically necessary" treatment is covered by third party insurance but other interventions are not, however desirable they may be because they improve competence. To derive these models they studied decisions made in the Harvard Community Health Plan, and HMO in the Boston area that offers unlimited coverage to persons "with severe psychiatric and substance abuse disorders" and decisions made in University of Toronto teaching hospitals where, under the Canadian health care system, physicians have discretion as to which interventions are offered to whom. When they studied case material and questioned the mental health professionals involved as to why certain services were or were not provided, they discovered that within the same clinical setting there was considerable variation among the providers in what they considered "necessary" and therefore eligible for intervention without cost to the client. A meta-analysis of the cases led to the conclusion that three models of reasoning seemed to underlie the actions of providers, though of course it is unlikely that any of them articulated them as such to themselves.

What Sabin and Daniels (1994) called the *Normal Function* model assumes that "the central purpose of health care is to maintain, restore or compensate for the restricted opportunity and loss of function caused by disease and disability" (Sabin & Daniels, 1994). The legitimate objective is to "restore people to the range of capabilities they would have had without the pathological condition" and to prevent further deterioration. A distinction is made, by those whose decisions are structured by this model, between people whose conditions can be diagnosed as mental disorders (DSM-IV) and those who are "less fortunate" in the natural and social lotteries that allocate personality strengths and weaknesses.

Their second model is called the *Capability* model and is more generous, more inclusive. The goal of interventions should go beyond restoring people to the pre-pathological conditions to helping them enhance personal capability. The assumption here is that we have, and should apply, techniques that will enhance competence — increase sociability, confidence, adaptability, etc. The aim is to help those with diminished capability, whether by inheritance or because of life experience, to allow them to become equal competitors in life. This model accepts as eligible for intervention any impairment of functioning that is amenable to improvement even if it is not a direct symptom of a DSM-IV mental disorder.

The third, the *Welfare* model is more expansive still. The assumption here is that eligibility for treatment should be extended to those who are at a social disadvantage because of "attitudes or behaviour patterns they did not choose to develop and are not independently able to alter or overcome." This model permits the mental health professional to treat virtually anyone who is suffering, not only because of a mental disorder or because of diminished capability in some desirable personality attribute but also because difficult life challenges are causing pain. Those mental health professionals who operate most comfortably under this model find it morally repugnant to deny help to people whose competence in life can be improved by therapy.

Which of these models is the right one? Apart from which one reflects most accurately the importance of environmental relative to person factors for the development and maintenance of intellectual, emotional and social health, which is ethically superior? Justice requires that equals be treated equally and unequals be treated unequally in accordance to their inequality, to paraphrase Kant. An in-depth examination of these models according to various theories of distributive justice is beyond the scope of this chapter. Scarce resources, such as money for health care, therapist time, and educational programs can be allocated, for example according to principles of first-come–first-served, degree of need, degree of opportunity, likelihood of success of interventions, "worthiness," etc.

In the real world, of course, not only client-dependent factors and ethical considerations guide clinician and administrator behaviour. Clinicians prefer certain interventions over others and tend to offer clients what they prefer to do and feel comfortable in doing, though they have an ethical responsibility to use where possible treatments of demonstrated effectiveness (e.g., Canadian Psychological Association, 1991; Chambless et al., 1996). Administrators are most sensitive to budgetary and administrative constraints. No doubt, administrators and taxpayers would insist on the more manageable *Normal Function model* which makes easier a distinction between treatment that is necessary and enhancement that is merely desirable. Many mental health providers, educators, and sophisticated would-be clients prefer the more expansive models.

The foregoing has been a broad examination of resource allocation for competence improvements, using primarily examples from mental health care to try to illustrate the issues. As this book demonstrates, there are important determinants of competence that lie outside the health field — in the social environment of infants and pre-school children, the educational system, the interpersonal climate of the family, the cultural and physical environment, employment opportunities, opportunities for personal development of adults, maintenance of competence of the elderly, etc. Even though in most of these areas there is less clearly defined social entitlement than there is in health care, we believe that the same issues arise.

Is the enhancement of human competence a desirable end, for the individuals themselves and for society? Of course the answer is "yes." Is it ethical to curtail unnecessarily the improvement of competence? Of course the answer is "no." Must cost be considered? When societal and personal resources are limited and many desirable claims are made on those resources, cost cannot be disregarded. Some of the considerations relevant in considering cost and entitlement have been touched on in this chapter. This remains an area of challenge to interdisciplinary scholarship and reflection.

REFERENCES

Anisfeld, E., Casper, V., Nozyce, M., & Cunningham, N. (1990). Does infant carrying promote attachment? An experimental study of the effects of increased physical contact on the development of attachment. *Child Development, 61*, 1617–1627.

Canadian Psychological Association (1991). *Canadian code of ethics for psychologists*. Ottawa: Canadian Psychological Association.

Chambless, L., Sanderson, W., Shoham, V., Johnson, S., Pope, K., Crits-Christoph, P., Baker, M., Johnson, B., Woody, S., Sue, S., Beutler, L., Williams, D., & McCurry, S. (1996). An update on empirically validated therapies. *Journal of Consulting and Clinical Psychology, 48*, 3–23.

Kramer, P.D. (1993). *Listening to Prozac*. New York: Viking Press.

Mrazek, P., & Haggerty, R. (1994). *Reducing risks for mental disorders: Frontiers for preventive intervention research: Summary*. Washington, D.C. National Academy Press.

Pierson, D., Walker, D., & Tivnan, T. (1984). A school-based program from infancy to kindergarten for children and their parents. *The Personnel and Guidance Journal, 62*, 448–455.

Sabin, J. E., & Daniels, N. (1994). Determining "Medical Necessity" in Mental Health Practice. *Hastings Centre Report, 24*, 5–13.

Schweinhart, L. J., & Weikart, D. (1988). The high/Scope Perry Preschool Program. In R. Price, E. Cowen, R. Lorion, & J. Ramos-McKay (Eds.), *14 ounces of prevention: A casebook for practitioners* (pp. 53–65). Washington: American Psychological Association.

Slomka, J. (1992). Playing with Propranolol. *Hastings Centre Report, 22*, 13–17.

INDEX

Academic competence/performance, 10, 182
Acceptance, 15
Active cognitive stimulation, 162
Activity involvement, 182
Acute Lymphoblastic Leukemia (ALL), 12, 17
Adaptation, 27, 38
    to cancer, integrated model of, 15–16
Adolescent alcohol use/abuse, 3, 7, 101–113
    cohort (historical) differences in, 104–105
    correlates and causes of, 106–110
    definition and measurement of, 102–103
    demographics of, 107–108
    personal characteristics and, 108–109
    prevalence of, 103–106
    prevention programs and, 110–113
Adolescents
    alcohol use/abuse in: see Adolescent alcohol
        use/abuse
    social competence in, 95–98
Adults
    cancer in, 12–16
    prime phase of life in, 133–138
Adversity, 10, 38
Advertising of alcohol, 111–112
Affect, 33–34
Affect attunement, 42
Affective intersubjectivity, 44
Affect regulation: see Emotional (affect) regulation
African Americans
    adolescent alcohol use/abuse in, 105–106, 108
    in Democratic model, 181
    poverty in, 72
Age
    adolescent alcohol use/abuse and, 104, 107
    cancer and, 13–14, 15, 16
Aggression, 35
    adolescent alcohol use/abuse and, 109, 110
    day care and, 80–81
    in maltreated children, 34
    popularity and, 96
Aging, stereotypes of, 3, 142–144, 147

Ainsworth Sensitivity Scales, 42
Alcohol abuse
    in adolescents: see Adolescent alcohol use/abuse
    defined, 102
Alcohol dependence, defined, 102
Alcoholism, defined, 102
American Drug and Alcohol Survey, 105
Anagram task, 127
Anger, 28, 65, 66, 67
    affective quality of parent–child relationship and, 68
    inappropriate coping with, 63
    risk-taking and, 125, 126
Antisocial personality, 108
Anxiety
    adolescent alcohol use/abuse and, 108, 109
    cancer and, 13, 14, 16, 21, 22
Asian Americans, 105, 108
Assessment
    of cognitive competence in infants, 45–50
    of competence in infants, 39–40
    of socioemotional competence in infants, 40–45
Attachment, 28
    to day care educators, 86–87
    in prime adulthood, 136, 137
    socioemotional competence and, 67–68
    strange situation assessment of, 42
Attention, 46–47
    joint, 48–49, 50
    self-, 31
    visual, 47
Autism, 49
Autobiographical narratives, 124
Avoidance, 14–15, 16, 18–19, 21, 63

Bayley Scales of Infant Development (BSID-II), 43, 48
Bayley Scales of Mental and Motor Intelligence, 45
Beer, 103, 107, 109
Behavioral competence, 10
Behavioral tilt, 136
Behaviorism, 94
Behavior Rating Scales (BRS), of BSID-II, 43

Binge drinking, 102, 103
Blacks: *see* African Americans
Blood alcohol concentration (BAC), 104, 111
Bone marrow aspirations, 12, 20
Breast cancer, 11–16
Brookline Early Education Project, 195–196
BSID-II: *see* Bayley Scales of Infant Development

Canada Health Act, 197
Cancer, 2, 9–23
    in adults, 12–16
    breast, 11–16
    in children, 12, 17–21
    demographic characteristics of patients, 13–14,
        15–16
    family coping with, 21–22
    integrated model of successful adaptation to, 15–16
    lung, 11
    optimism and, 14, 15, 16
    prostate, 11
    social support and, 14, 20
    as a stressor, 11–12, 13, 19
Capability model, 198
CARE-INDEX: *see* Child–Adult Relationship Experi-
    mental Index
Castaneda, Carlos, 197
Category-driven research, 136–137
CBCL: *see* Child Behavior Checklist
CHAT: *see* Checklist for Autism in Toddlers
Checklist for Autism in Toddlers (CHAT), 49
Chemotherapy, 12, 17, 18, 19
Child–Adult Relationship Experimental Index (CARE-
    INDEX), 43
Child Behavior Checklist (CBCL), 17
Children
    affective quality of parents' relationship with, 67–68
    cancer in, 12, 17–21
    day care and: *see* Day care
    discussion of emotion and, 68–69, 82–83
    emotional development in: *see* Emotional develop-
        ment
    expression of emotion in, 65–67, 82
    maltreatment of, 34
    parental reaction to emotions in, 61–65, 82
    political models of development and: *see* Political
        models
    social competence in, 95–98
    socioemotional competence in, 2, 59–74
Choice manipulation, 129
Cirrhosis, 106
Client effects, 4–5
Clinical depression, 164
Cocaine abuse, 42, 108
Cognitive competence, *see also* Cognitive development
    complaints about, 162
    in infants, 2, 45–51
    in late life, 159–164
Cognitive development, 19–20; *see also* Cognitive
    competence

Cognitive reframing, 169
Cognitive restructuring, 16, 20
Cohort effects
    in adolescent alcohol use/abuse, 104–105
    in late life, 144
Communication Predicament of Aging model,
    143
Community, 181–182, 183
Compensation, 3, 142, 148, 150
Compete, etymological analysis of, 92–93
Competence predicament, 142–144
Computer skills, 147, 152–153
Conceptualizations of competence, 10–11
Conditioning paradigm, 46
Conduct disorder, 108
Conflict
    in depressed elderly, 165
    in families, 71–73
    in friendships, 97–98
    marital, 71–72
Consciousness, 28–30
Context-specific processes, 135, 136
Control, perceived, 15, 153
Convoy model, 167
Coping
    in adult cancer patients, 12–16
    as enacted competence, 10–11
    in families of cancer patients, 21–22
    in pediatric cancer patients, 17–21
    in prime adulthood, 137–138
Cost considerations, 193–199
Crime, 118
Cross-modal transfer, 46
Cystic fibrosis, 20

Day care, 2–3, 79–88
    educator warmth and harshness in, 87, 88
    home care vs., 2, 79, 80–81
    lessons from maternal socialization in, 84–86
    role of educator in socioemotional development,
        86–88
    staff turnover in, 87, 88
    teacher–child ratio in, 87
Deferred-imitation paradigm, 48, 49
Definitions of competence, 92–93, 94
Dementia, 162–163, 170
Democratic model, 177–185
    Republican model vs., 188–191
    risk factors in: *see* Risk factors
Demographics
    in adolescent alcohol use/abuse, 107–108
    of adult cancer patients, 13–14, 15–16
Denial, 15, 18–19
Dependence-support scripts, 153
Depletion syndrome, 164
Depression
    adolescent alcohol use/abuse and, 108, 109
    cancer and, 13, 14, 16, 21, 22
    clinical, 164

Depression (*cont.*)
  effect of maternal depression on infants, 41
  in late life, 164–166
  in maltreated children, 34
  minor, 164
  primary, 164
  secondary, 164
  subclinical, 164, 165
Development
  cognitive, 19–20
  emotional, 2, 28–35
  in infants, 37–38
  political models of: *see* Political models
  problems and delays in, 45–46, 47–48
  scientific models of, 177, 189
Developmental tasks, 134–135
Diagnostic and Statistical Manual of Mental
    Disorders (DSM-IV), 102, 164, 195, 196,
    198
Discussion of emotions, 68–69, 82–83
Disengagement, 15, 16
Disgust, 28
Dissing, 34
Distraction, 11, 18–19, 20
Divorce, 71, 72, 73
Down syndrome, 47, 48, 49
Driving
  drunk, 102, 105, 110, 111
  in late life, 150
Dyads, 3, 39, 41, 43, 95, 97–98
Dysthymia, 164

Early Childhood Environment Rating Scale, 81
Early Secondary Intervention Program, 112
EAS: *see* Emotional Availability Scales
East Asian countries, 108
Economic stress, 72, 73
Education level, 13, 14, 15–16
Ego threat, 120–121, 130
Elderly: *see* Late life
Embarrassment, 2, 30–31, 43, 125
Emotional Availability Scales (EAS), 42–43
Emotional behavior, 35
Emotional competence, 2, 27–28, 35; *see also* Emo-
    tional development; Socioemotional compe-
    tence
  in late life, 164–167
Emotional control, 137
Emotional development, 2, 28–35; *see also* Emotional
    competence
Emotional distress, 123–126, 130; *see also* Personal
    distress
Emotional reactivity, 136
Emotional (affect) regulation, 20, 118, 128, 137
Emotion-focused reactions, 16, 63
Emotions
  child effects on parental view of, 69–70
  discussion of, 68–69, 82–83
  expression of in prime adulthood, 136–137

Emotions (*cont.*)
  link between parents' and children's expression of,
    65–67, 82
  parental reactions to children's, 61–65, 82
  primary, 28, 35
  reaction of day care educators to, 88
  self-conscious, 2, 31, 32–34, 35
Empathy, 31, 60–61, 62, 66, 82
  affective quality of parent–child relationship and, 68
  defined, 60
  discussion of emotion and, 69
  training in, 73
Engagement, 16
Entitlement, 194
Environment, 38, 92
  conditions in as risk factors, 179
  continuity of risk in, 180
  physical, 152–153
  in political models: *see* Democratic model; Republi-
    can model
  social: *see* Social environment
Erikson, Erik, 3, 93, 136
Ethics of resource allocation, 4, 193–199
Ethnicity: *see* Race/ethnicity
Etymological analysis of competence, 92–93
Evolutionary theory, 27
Executive function, 129
Exercise, 152
Expected gain, 124
Experience-dependent brain plasticity, 136
Experience-expectant brain plasticity, 136

Face-name strategy, 151
Face-to-face interactions, 41–42
Fagan Test of Infant Intelligence, 47
Failure
  overshooting and, 120
  positive and negative language and affect as conse-
    quence of, 33–34
  socialization of responsibility for, 32
Failure-to-thrive infants, 47
Familiarization-novelty paradigm, 46
Families
  adolescent alcohol use/abuse prevention and,
    112–113
  of cancer patients, 21–22
  conflict in, 71–73
  socioemotional competence and, 71–73
Family management of the community, 181, 182
Family process, 181, 182
Family structure, 181, 182
Fathers, 68, 69
Fear, 28
Feminist psychology, 98
Field's Interaction Rating Scales, 41
Fluid intelligence, 161
Friendly PEERsuasion, 112
Friendship, 97–98
Frontal lobe damage, 30

Gender differences
    in adolescent alcohol use/abuse, 103, 105–106,
        107–108
    in day care effects, 80
    in discussion of emotion, 69
    in embarrassment, 30
    in maltreated children, 34
    parental reactions to emotions and, 64
    in shame, 32
Gender role theory, 98
Global self-attention, 31
GLOS: see Greenspan–Lieberman Observation
    System
Goals, setting and reaching, 118, 119–121
Grant Foundation, 38
Gratification delay, 119
Greenspan–Lieberman Observation System (GLOS),
    43
Groups, 3, 39, 95, 98
Growth hormone, 194–195, 196
Guilt, 2, 31, 43, 68

Habituation–dishabituation paradigm, 46
Handgrip task, 127, 129
Harvard Community Health Plan, 198
Head Start, 39, 40
Hispanics, 105, 108
Historical development of competence concept, 3,
    91–99; see also Erikson, Erik; Piaget, Jean;
    White, Robert
Home care, 2, 79, 80–81
Homicide, 105
Human potential, 194
Humor, 15
Huxley, Aldous, 197

IFEEL pictures, 43–44
Imitation, 48, 49, 50
Impulse control, 118
Indians: see Native Americans
Individual, 3, 39, 95, 96–97
Infants, 37–53
    assessment of competence in, 39–40
    cognitive competence in, 2, 45–51
    competence in, 37–39
    failure-to-thrive, 47
    interventions for, 39–40, 50–53
    low-birth-weight, 46, 47
    preterm, 42, 47
    self-recognition in, 29–30
    socioemotional competence in, 2, 40–45, 51
Information processing, 46, 47–48
Institute of Medicine, 166
Instrumental problem-solving, 63
Intelligence (IQ), 45, 179, 180, 186–187
Intelligence (IQ) tests, 45, 179
Intensity, 136
Interactions, 97–98
International Classification of Diseases (ICD), 196

Interventions
    for cognitive competence in late life, 160–163
    for improving competence in late life, 151–153
    inadvertent effects of, 5
    for infants, 39–40, 50–53
    for socioemotional competence, 73–74
Intrusive thoughts, 11, 13, 14–15, 16, 21
IQ: see Intelligence (IQ)

James, William, 37
Johnson, Ben, 196
Joint attention, 48–49, 50
Joy, 28

Keyword strategy, 151
Kramer, Peter, 197

Labeling/explaining, 84
Language, 51
    focus of attention and, 49
    positive and negative as consequence of success or
        failure, 33–34
Late life, 3, 7, 141–154, 159–171; see also Successful
    aging
    cognitive competence in, 159–164
    cohort effects in, 144
    competence predicament in, 142–144
    core assumptions about, 146
    emotional competence in, 164–167
    interventions toward improving competence in,
        151–153
    social competence in, 169–170
    well-being in, 166–169
Latency phase, 93
Latin American countries, 106, 108
Learned helplessness, 143
Leary, Timothy, 197
Leukemia, 19
    acute lymphoblastic (ALL), 12, 17
Life Skills Training program, 112
Limit-testing procedures, 135
Listening to Prozac (Kramer), 197
Local knowledge, 97
Longitudinal studies
    of adolescent alcohol use/abuse, 104, 105, 107, 110
    of cancer patients, 13
    of day care, 80
    of prime adulthood, 136
    of procrastination, 122
    Rochester, 178, 179, 180–181, 182, 186
    of self-regulation, 118–119, 128
    of well-being, 166
Lottery experiment, 124–125, 126
Low-birth-weight infants, 46, 47
Lumbar punctures, 12, 17, 20
Lung cancer, 11

Major depression, 164, 165
Maltreatment of children, 34

Marital conflict, 71–72
Mechanistic models, 189
Medically necessary treatment, 196, 198
Memory, 143, 144, 163–164
  compensation and, 150
  complaints about, 162
  dementia and, 162–163
  interventions for improving, 151–152, 160–161, 162
  selection and, 149
  verbal, 147
  working, 147
Memory wallets, 162–163
Mental retardation, 30
Meta-analysis
  of cognitive competence, 47
  of depression in late life, 164
  of friendship, 97–98
  of health care system, 198
  of memory interventions, 151, 161, 162
  of popularity, 96
  of well-being, 167
Method of loci, 151
Mexican Americans, 105, 108
Minor depression, 164
Mirror recognition, 28–30
Mnemonic strategies, 149, 150, 151, 162
Mother–Child Home Project, 39
Mothers
  affective quality of child's relationship with, 67–68
  discussion of emotion and, 68–69, 82–83
  expression of emotion in, 66, 67, 82
  perceptions of child's emotional reactions in, 69–70
  reactions to children's emotions in, 62–65, 82
  socioemotional competence and, 41–43, 81–86
Mothers' Project, 43
"Motivation reconsidered: The concept of competence" (White), 92, 94

National health insurance, 195
National Household Survey on Drug Abuse, 103, 108
National Seniors Survey, 103–105
Native Americans, 105, 108, 109
Natural killer cells, 14
NBAS: *see* Neonatal Behavior Assessment Scale
Neglected children, 96, 97
Neonatal Behavior Assessment Scale (NBAS), 43
Normal Function model, 198, 199
Nursing homes, 143, 153

Objective self-awareness, 28; *see also* Consciousness
Odds-ratio analysis, 183–185
Ontogenetic recapitulation of phylogeny, 134
Optimism, 14, 15, 16
Optimization, 3, 142, 148, 149–150
Organismic models, 189
Other, 98
Overconfidence, 120–121, 130
Overshooting, 120

PACT: *see* Parent–Child Affect Communication Task
Parent authority styles, 137
Parent characteristics, 181, 182
Parent–Child Affect Communication Task (PACT), 83–84
Parent–Child Early Relational Assessment (PCERA), 43
Parents, *see also* Fathers; Mothers
  affective quality of child's relationship with, 67–68
  cancer in, 21–22
  cancer in children and, 19
  discussion of emotion and, 68–69, 82–83
  expression of emotion in, 65–67, 82
  perceptions of children's emotional reactions, 69–70
  reactions to children's emotions in, 61–65, 82
  standards set by, 32
Parents and Students Against Drinking and Driving, 105
PATHS curricula, 73–74
PCERA: *see* Parent–Child Early Relational Assessment
Peer relations
  adolescent alcohol use/abuse and, 109–110
  day care and, 80, 81, 85
  in Democratic model, 181, 182, 183
  social competence and, 95–98, 99
Peg-word method, 151
Performance focus, 32
Perry Preschool Program, 195–196
Personal distress, 61, 62, 64, 66; *see also* Emotional distress
Personality, 44
Personality theory, 98
Personal pronoun usage, 29, 30
Pessimism, 14, 16
Philadelphia study, 181–185, 186, 187, 189
Physical contextual features, 135
Physical environment, 152–153
Piaget, Jean, 3, 37, 48, 93–94
Plasticity, 136, 148, 150
"Playing with Propranolol" (Slomka), 197
Political models, 3–4, 177–192; *see also* Democratic model; Republican model
Popularity, 63, 64, 96
Post Traumatic Stress Disorder (PTSD), 14, 19
Poverty, 72, 73
Practice, 147–148, 150
Preferential looking paradigm, 46
Prenatal/Early Infancy Project, 39
Pretend play, 29, 49
Preterm infants, 42, 47
Prevention of Mental Disorders (report), 166
Pride, 2, 28, 31–32, 33–34, 43, 68
Primary depression, 164
Primary emotions, 28, 35
Prime adulthood, 133–138
Problem behavior, 182
Problem-focused reactions, 16, 63–64
Problem solving, 16, 20

Problem-Solving Communication, 112
Process-oriented studies, 136–137
Process specificity, 135
Procrastination, 122–123, 130
Project CARE, 39
Propranolol, 196–197
Prostate cancer, 11
Psychological adjustment, 182
    to pediatric cancer, 17–18
Psychological capital, 185
Puerto Ricans, 181
Punishment, 33
Pygmalion potential, 136

Race/ethnicity, see also specific racial and ethnic groups
    adolescent alcohol use/abuse and, 105–106, 108
    in Democratic model, 181
Radiation therapy, 17, 18
Recovered label, 136
Regulation
    self-: see Self-regulation
    social, 7
Rejected children, 96–97
Relationships, 97
Remedial strategies, 135
Reminiscence therapies, 165
Repression, 18–19
Republican model, 177–178, 185–188
    Democratic model vs., 188–191
Reserve capacity, 147, 148, 160, 163
Resignation, 20
Resilience, 10, 38, 136
Resource allocation, 4, 193–199
Risk factors
    accumulating, 179
    continuity of environmental, 180
    environmental conditions as, 179
    identification of, 182–183
    scores for, 183
Risk-taking, 123–126, 130
Rochester Longitudinal Study, 178, 179, 180–181,
        182, 186
Role theory, 167
Romania, 38
Ruminative coping, 22

Sadness, 28, 125
Scaffolding, 84, 88
Schools, alcohol prevention programs in, 112–113
Scientific models, 177, 189; see also Political models
Secondary depression, 164
Secure attachment: see Attachment
Selection, 3, 142, 148, 149
Selective serotonin re-uptake inhibitors (SSRIs), 197
Self, 98
Self-attention, 31
Self-competence, 182
Self-conscious emotions, 2, 31, 32–34, 35
Self-criticism, 16

Self-defeating (destructive) behavior, 118, 123–126, 130
Self-esteem, 120–121, 130
Self-focus, 32, 33–34
Self-knowledge, 120, 121
Self-management, 121, 130
Self-recognition, 28–31
Self-regulation, 3, 7, 117–130
    benefits of, 117–119
    defined, 117
    emotional (affect), 20, 85, 86, 118, 128, 137
    frontal lobe damage and, 30
    nature of, 126–130
    procrastination and, 122–123, 130
    self-defeating behavior and, 118, 123–126, 130
    in setting and reaching goals, 118, 119–121
    socioemotional competence and, 59–60, 61, 62, 64
Sensation-Seeking Scale, 109
Sensory deprivation, 166
Separation, 85–86, 137
Shame, 2, 31–32, 33–34, 35, 43, 68
Sibling caretaking, 83–84
Significant event technique, 162
Single-parent families, 72, 180, 189–190
Sleeper pattern, 136
Social activities, 168
Social cognition, 97
Social competence, 10, 95–98; see also Socioemotional competence
    in late life, 169–170
    in Republican model, 187
Social environment
    adolescent alcohol use/abuse and, 106–107, 109–110
    in late life, 152–153
Social integration, 168–169
Socialization
    maltreatment in, 34
    of self-conscious emotions, 32–34, 35
    of socioemotional competence, 59–74, 79–88; see also Day care; Socioemotional competence
Social phobia, 196
Social regulation, 7
Social roles, 168
Social signalling, 137
Social skills, 63
Social skills training, 165, 169–170
Social support, 14, 20
Socioeconomic status (SES), 47, 178, 179, 181, 189–190
Socioemotional competence
    affective quality of parent–child relationship and, 67–68
    assessment of, 40–45
    child effects on, 69–70
    in children, 2, 59–74
    day care and: see Day care
    definitional issues in, 60–61
    discussion of emotion and, 68–69
    in infants, 2, 40–45, 51
    interventions for, 73–74
    maternal socialization of, 41–43, 81–86

Socioemotional competence (*cont.*)
  parent and child expressivity in, 65–67
  parent reactions to child's emotions and, 61–65
  in prime adulthood, 136
Socioemotional selectivity theory, 167
Sons of alcoholics, 107, 108–109
Sows' ear principle, 136
Spaced retrieval training, 162
Specialization, 146
Specificity, 135
Specific self-attention, 31
Stabilities in competence, 5–6
Staff turnover, in day care facilities, 87, 88
Stage salient tasks, 44–45
Standard-Transformation-Return (STR) procedure, 47
Stanford–Binet test, 48
Stereotypes of aging, 3, 142–144, 147
Still-face (SF) procedure, 42
Strange situation, 42
Stress, 38; *see also* Coping
  cancer as a cause of, 11–12, 13, 19
  economic, 72, 73
  in prime adulthood, 137–138
Subclinical depression, 164, 165
Subjective value theory, 125–126
Success
  positive and negative language and affect as consequence of, 33–34
  self-regulation and: *see* Self-regulation
  socialization of responsibility for, 32
Successful aging, 142, 144–151
  antecedents of, 147–148
  basic premises of, 146–147
  defining outcomes in, 150–151
Suicide, 105, 123–124
Sympathy, 60–61, 62, 64, 66, 68, 69, 70

"Tale, The," 141–142, 151, 154
Task focus, 32, 33–34
Temperament, 11, 30, 35, 44, 60, 137
Temptation, resisting, 127–128
Thought suppression, 11
Thought suppression task, 127, 128, 129
Touch, 42
Transactional models, 52
Type A personality, 122
Type B personality, 122
Type I alcoholism, 109
Type II alcoholism, 109
Typing skills, 147, 150

Undershooting, 120
University of Toronto, 198

Valence, 136
Venipunctures, 12, 17
Verbal memory, 147
Violence, 118
Visual attention, 47
Volunteering, 168

Welfare model, 198
Well-being, 166–169
White, Robert, 3, 92, 93–95, 99n2
Whites
  adolescent alcohol use/abuse in, 105, 106, 107, 108, 109, 110
  in Democratic model, 181
Wishful thinking, 16, 20
Working memory, 147
Working models, 137

Zero to Three, 39

DATE DUE

MAY 2 3 2009		

GAYLORD   #3523PI   Printed in USA